Critical Ecolinguistics Within Singapore, ‘the Garden City’

Bloomsbury Advances in Ecolinguistics

Bloomsbury Advances in Ecolinguistics emerges at a time when businesses, universities, national governments and many other organisations are declaring an ecological emergency. With climate change and biodiversity loss diminishing the ability of the Earth to support life, business leaders, politicians and academics are asking how their work can contribute to efforts to preserve the ecosystems that life depends on.

This book series explores the role that linguistics can play in addressing the great challenges faced by humanity and countless other species. Although significant advances have been made in addressing social issues such as racism, sexism and social justice, linguistics has typically focused on oppression in human communities and overlooked other species and the wider ecosystems that support life. This is despite the disproportionate impact of ecological destruction on oppressed groups. In contrast, this book series treats language as an intrinsic part of both human societies and wider ecosystems. It explores the role that different areas of linguistic enquiry, such as discourse analysis, corpus linguistics, language diversity and cognitive linguistics can play at a time of ecological emergency.

The titles explore themes such as the stories that underpin unequal and unsustainable industrial societies; language contact and how linguistic imperialism threatens the ecological wisdom embedded in endangered languages; the use of linguistic analysis in ecocriticism, ecopsychology and other ecological humanities and social sciences; and emerging theoretical frameworks such as Harmonious Discourse Analysis. The titles also look to cultures around the world for inspirational forms of language that can lead to new stories to live by. In this way, the series contributes to linguistic theory by placing language fully in its social and ecological context, and to practical action by describing the role that linguistics can play in addressing ecological issues.

Titles published in the series:

Corpus-Assisted Ecolinguistics, Robert Poole

Ecolinguistics and Environment in Education, Emile Bellewes

Ecological Communication and Ecoliteracy, edited by Maria Bortoluzzi and Elisabetta Zurru

Exploring Ecolinguistics, Douglas Mark Ponton

Language and Ecology in Southern and Eastern Arabia, edited by Janet C.E. Watson, Jon C. Lovett and Roberta Morano

Language as an Ecological Phenomenon, edited by Sune Vork Steffensen, Stephen Cowley and Martin Döring

Multispecies Discourse Analysis, Gavin Lamb

Storytelling and Ecology, Anthony Nanson

TESOL and Sustainability, edited by Jason Goulah and John Katunich

Critical Ecolinguistics Within Singapore, 'the Garden City'

Edited by
Todd LeVasseur

BLOOMSBURY ACADEMIC
LONDON • NEW YORK • OXFORD • NEW DELHI • SYDNEY

BLOOMSBURY ACADEMIC
Bloomsbury Publishing Plc, 50 Bedford Square, London, WC1B 3DP, UK
Bloomsbury Publishing Inc, 1359 Broadway, New York, NY 10018, USA
Bloomsbury Publishing Ireland, 29 Earlsfort Terrace, Dublin 2, D02 AY28, Ireland

BLOOMSBURY, BLOOMSBURY ACADEMIC and the Diana logo are trademarks of Bloomsbury Publishing Plc

First published in Great Britain 2026

Series design: Ben Anslow

A catalogue record for this book is available from the British Library.

Library of Congress Cataloging-in-Publication Data
Names: LeVasseur, Todd editor Title: Critical ecolinguistics within Singapore, 'the Garden City' / edited by Todd LeVasseur. Description: New York : Bloomsbury Academic, 2026. | Series: Bloomsbury advances in ecolinguistics | Includes bibliographical references and index. | Summary: ""This book offers a critical ecolinguistics case study of Singapore, and of the stories told about and by it in relation to its efforts of sustainability, response to climate change, and commitment to ecomodernity. The authors offer insights into contemporary issues around sustainable urbanism, the role of discourses in conceptualizing and advocating for sustainability, methods on how to apply an ecolinguistical lens to various 'texts' that constitute a nation city, and an understanding of how various discourses may promote or impede climate resiliency and adaptation""– Provided by publisher.
Identifiers: LCCN 2025030094 (print) | LCCN 2025030095 (ebook) |
ISBN 9781350438491 hardback | ISBN 9781350438507 paperback |
ISBN 9781350438521 epub | ISBN 9781350438514 pdf
Subjects: LCSH: Ecolinguistics–Singapore
Classification: LCC P39.5.C75 2026 (print) | LCC P39.5.C75 (ebook)
LC record available at https://lccn.loc.gov/2025030094
LC ebook record available at https://lccn.loc.gov/2025030095

ISBN: HB: 978-1-3504-3849-1
ePDF: 978-1-3504-3851-4
eBook: 978-1-3504-3852-1

Series: Bloomsbury Advances in Ecolinguistics

Typeset by Newgen KnowledgeWorks Pvt. Ltd., Chennai, India

For product safety related questions contact productsafety@bloomsbury.com.

To find out more about our authors and books visit www.bloomsbury.com and sign up for our newsletters.

We dedicate this book to the humans, fauna, flora, fungi and waters of Singapore; to the country's continued flourishing; and to Southeast Asia's ability to thrive by applying ongoing decolonial praxis despite the changes of climate to come. Todd specifically dedicates this book to the youth climate leaders of Southeast Asia and of Singapore, specifically of whom this book's co-authors represent the best the region has to offer to Gaia.

Contents

Figures

Contributors

Beverley Jia Ying Choo (she/they) graduated in 2023 from Yale-NUS College, Singapore, with a B.A. in Environmental Studies. She has a background in youth climate movements, environmental nonprofits and climate philanthropy and is deeply passionate about climate justice and marine conservation. Specifically, Beverley is interested in decolonial approaches to tackling our sociopolitical and environmental problems (everything is interconnected!) and building intercultural and interregional solidarity to achieve climate justice. In line with this, they co-founded GreenCheck, a youth-led nonprofit that empowers youth to take action for climate justice and redirects resources to community-led climate action groups in Most Affected Peoples and Areas. Beverley believes strongly in the power of community and kinship – between humans and other non-human inhabitants of our world – to bring about the change needed to create a more regenerative and equitable future.

Todd LeVasseur (he/him) holds a PhD in the study of religion and nature (University of Florida, United States) and an M.Sc. from the Centre for Human Ecology in Scotland. He is a senior instructor in the Environmental and Sustainability Studies Program at the College of Charleston, SC, United States, where at the time of writing, he is also Assistant Director of the same program. He has published widely on issues related to queer ecologies, sustainability in higher education, sustainable agriculture, climate change, animal humanities, ecolinguistics and religion and nature/ecology. From 2022 to 2024, he was Senior Lecturer in the Environmental Studies programme at Yale-NUS College, Singapore.

Lecarl Lim (he/him) is a 2024 Environmental Studies (B.A.) graduate from Yale-NUS College, Singapore, with a minor in Urban Studies. He has a little white dog called Snoopy and a slightly larger one called Julius, and they taught him to love fellow animals and deepened his conviction in the equal value of human and more-than-human lives. Inspired by the Ubuntu wisdom, 'I am because we are', he has dedicated much effort to building communities to navigate our socio-ecological crisis. Previously, he co-chaired the Inter-University Environmental Coalition of Singapore – a youth-led community of collaboration and policy participation. He later co-founded Post-Growth Singapore, a ground-up movement dedicated to envisioning and building a future beyond the reliance on economic growth as the sole path to prosperity, and one focused on fostering a socially just, equitable society in harmony with nature instead. Lecarl envisions a future where our whole of society comes together to recentre universal care for fellow humans, more-than-humans and future generations.

Kristina Gweneth Ponce Simundo, or Kristina Gweneth (they/she), graduated in 2024 from Yale-NUS College, Singapore, with a B.A. in Environmental Studies and a minor in Arts and Humanities. Born in the Philippines but having grown up in Singapore, Kristina holds a lot of love for the country their family migrated to and the city's unique spaces and narratives. This love pushes them to have bigger hopes for Singapore to be in stronger climate solidarity both regionally and globally. When Kristina first encountered Ecolinguistics in their Environmental Humanities undergraduate class, it provided a powerful tool to identify and therefore more effectively resist unequal, violent and oppressive power dynamics in our ecosystem – especially interrogating the intersections between climate change and marginal identities of race, class, gender and sexuality. She hopes that this book enables readers to identify their own power over the discourses in their lives, especially our collective people power in rewriting the most important story of all – the story of our future in the face of the climate crisis.

Hannah Shueh Yi Ng (she/her) is a 2024 Environmental Studies (B.Sc.) graduate from Yale-NUS College, Singapore. She is a passionate lover of the ocean, whose hope has always been to make the world a better place. Her academic passions led her across continents to Tanzania and the United States of America to lean into as many academic fields as possible (though she focused on scientific, historical and social science intersections most), and she believes that the only way to answer climate change's wicked problems is through intersectional and interdisciplinary answers. Her love for the environment comes from a deep peace that she experiences underwater with the living ocean, a love for fellow peoples regardless of identity and her calling as a Roman Catholic to be a steward of the planet. Her formation in Ignatian spirituality has helped her to see God through the beauty in creation and a resonance with the idea of Radical Love – love that transcends any and all boundaries. It is both her great grief and hope that she can contribute to protecting the planet and all its living things through her work.

Anjali Venketram (she/her) is a 2024 graduate of Yale-NUS College, Singapore, who majored in Environmental Studies and minored in Urban Studies. Coming from India, settling in Singapore was a process where she learned a lot about what environmentalism could look like. While she continues to be impressed with environmental action in Singapore, her experiences also led her to critically reflect on the broader implications of environmentalism in a capital-driven society. She became deeply affected by the tensions that arise when the pursuit of economic growth takes precedence over the health and well-being of communities, including the non-human members of ecosystems. Anjali is inspired to explore ways to reconcile these competing priorities, fostering a vision of environmentalism that is equitable, inclusive and centred on the flourishing of all life forms.

Tan Yock Han, or Yock (he/him), graduated from Yale-NUS College, Singapore, in 2025 with a major in Environmental Studies (B.A.) and a minor in Urban Studies. Informed by his experiences growing up in hypermodern, cosmopolitan Singapore, a personal passion of his is investigating how spaces can facilitate healthier relationships between human and non-human nature. In order to erode away at the binarization of the two, he believes that the inherent value of nature should be appreciated before its instrumental values. A central tenet to his own personal ecosophy is striking an appropriate balance between urban development and natural and cultural conservation, and he advocates that this balance should be underpinned by traditional ecological knowledge systems. A key highlight of his climate journey was spending the first half of 2024 in Hawai'i, United States of America, where he immersed himself in discourses centred around sovereignty movements and the preservation of indigenous identities.

Preface

Todd LeVasseur

A key insight in the interdisciplinary field of ecolinguistics, and therefore one for this edited volume, is that humans live by stories (Stibbe 2015). Within this, ecolinguistics especially focuses on how these stories relate to the more-than-human world. Or as the environmental philosopher Max Oelschlaeger put it, humans are storytelling culture dwellers (Oelschlaeger 1994, 9), where such dwelling (Ingold 2022) always occurs in what I call 'ecologies of alive places' (LeVasseur 2021). This, then, is an edited compendium about certain place-based stories. Specifically, these are stories about dwelling in the living ecologies of Singapore, an island nation city in Southeast Asia that is home at the time of writing to approximately six million people.

The stories that constitute this book are:

1. stories about Singapore told by various Singaporeans;
2. stories about Singapore told by others outside of Singapore, as these stories especially relate to the assumed sustainability of Singapore;
3. thus stories about sustainability, and stories about what that concept means, especially in urban-island areas; and therefore,
4. there is an underlying story running throughout this book about anthropogenic climate change. This is because the unfolding of this biogeochemical earthen story will impact the other three stories throughout this century and beyond.

Sharing and analysing these stories emerges from a cohort of at that time undergraduate students who in the fall semester of 2022 matriculated into the Foundations of Environmental Humanities class I taught at Yale-NUS College (most often just called 'YNC'), Singapore's first liberal arts college co-founded in 2011 by the National University of Singapore (NUS) and Yale University. The mid-term for that course was to apply an ecolinguistic analysis according to the insights of Stibbe (2015) to any public places in Singapore. Specifically, students were tasked with finding an example of a *beneficial* discourse, *ambivalent* discourse and *destructive* discourse (to be discussed later). The results of these efforts were of such high quality that I approached the editors of this book series, presenting the idea of a project that would apply an ecolinguistical analysis to an entire city-nation, building on the insights and labour of the students for that class. This edited book is the result of that proposal. Key within the idea for this book is that the authors, outside of Todd, know Singapore as residents, either of YNC, which is a four-year residential college, or as being born and raised in Singapore. The latter group also lived on the Yale-NUS campus for four years, and all contributors have taken (or in my case, taught) a variety of Singapore – and Asia-focused environmental studies courses during their time at YNC. Therefore, the respective authors of this book speak from lived experience as experts on Singapore and its past, present and future.

One of the key realities of this book, and thus of living in Singapore, is that outside of me, all of the authors are also members of the climate generation. Consistent with global data on this generation, all authors feel a variety of climate emotions that many in the climate generation (as well as members of other generations, such as me) grapple with in their daily lives: anger, betrayal, grief, despair, apathy, hope, optimism and resistance (Ray 2020; Diffey et al. 2022; Hickman 2021). All authors are also cognizant of feelings and states of solastalgia (Albrecht 2007). This solastalgia is because the government of Singapore, despite its green reputation (on this, see the introduction and Chapter 1), is rapidly converting its green places and old historic neighbourhoods into high-rise apartments and gated communities. Almost all of the authors who grew up in Singapore have heard stories from aunties and uncles[1] who vividly remember fields and forests and community spaces where they used to play as children back in the 1900s. Some authors remember such fields and trees themselves. Yet these fields where they interacted with local flora and fauna have now been violently razed to make way for the Singapore of the future.

Readers should note that these climate emotions and lived experiences and hearing about how Singapore used to have more wild places all helped to inform the creation of this book. Further motivation stems from our collective desire to transition our species to more just, resilient and sustainable ways of living, including in Singapore. This transition also needs to occur in ways that allows for and supports the flourishing of non-human species and the ecosystems within which these other planetary kin (and of course, humans, too) dwell, including in Singapore.

Ecolinguistics and Discourses: An Overview

Arran Stibbe, one of the leading scholars of ecolinguistics and the co-editor of the series within which this book is published, states that ecolinguistics 'is about critiquing forms of language that contribute to ecological destruction, and aiding in the search for new forms of language that inspire people to protect the natural world' (Stibbe 2015, 1). He continues, explaining how 'the *stories-we-live-by*' (Stibbe 2015, 1) provide mental models that shape the actions and behaviours humans take. Or, as the religious studies scholar and scholar of peace and nonviolence Heather Eaton explains, stories are parts of a worldview. And key for her is that a worldview is an 'amalgam of the visions, ideas, ideals, and practices that interweave to produce cultural values, governance systems, and social identities. Worldviews are the way we imagine life and our lives together to produce the quality of our communities, the questions we raise, and the moral principles we choose' (Eaton 2017, 215). Eaton further explains that stories and worldviews are part of symbolic consciousness. Thus, the *stories-we-live-by* interact with this symbolic consciousness because humans always dwell in worldviews and stories, both. Therefore, the symbols and grammars of and in such stories shape human consciousness and thus shape how humans understand, and therefore respond to, interact with and engage with the world in its entirety. An iterative pattern emerges where stories help encode worldviews, and such worldviews shape the *stories-we-live-by*. These stories also reinforce, via socialization and repetition, the stories we tell and that we allow to shape our lives and behavioural choices.

I pause to note here that advertisements are a key form of storytelling. Their repetition, funded by hundreds of millions of dollars in advertising with the goal of coercing consumer purchases and brand identities, while making us think our main role in life is to shop, also wires our brains and the synapses within our brains. The same holds for the repetition of any and all *stories-we-live-by*,[2] meaning that human brains are shaped biophysically around thoughtways, and these thoughtways also shape our consciousness. Taken together, such stories, thoughtways and symbolic consciousness all interact in psychological, affective, emotional, social and religio-spiritual ways that inform our approach to reality and how we live within reality. In short, we are literally shaped by stories.

These stories also provide 'blueprints' and visions for desired futures. This is the case whether these futures perpetuate currently existing systems or whether the stories share worldviews and visions that challenge hegemonic systems.[3] The latter is especially true for ecolinguistics and the need to develop *beneficial* discourses that can articulate new stories by which to live in ecologies of alive places. Such stories are desperately needed during this era of biodiversity loss and changing weather and climate patterns, and with the rampant overconsumption by privileged humans of earthen ecosystems. Ecolinguistics builds on such ecological concerns and fuses them with the wide-ranging field of linguistics. This combination of eco and linguistics recognizes that how language is used and structured shapes how the group that uses that language sees and understands, and thus treats, the subjects and objects of that language.[4] The eco part, from the Greek *oikos*, which translates into house or home, posits that the language (texts, discourses and semiotics, whether verbal, written or material) people use about ecologies/ecosystems/environments, will thus shape their understanding and treatment of those biophysical, more-than-human places.

Furthermore, ecolinguistics focuses on how the *stories-we-live-by*, and the discourses, syntaxes and grammar of them as well as the worldviews and ideologies contained in them, coalesce into informing and shaping how humans understand the world. Ecolinguistics also focuses on how humans understand their place within biosocial worlds as described in those stories. More specifically, ecolinguistics investigates how the discourses, texts, grammars, semiotics and the *stories-we-live-by* depict and frame the environment and more-than-human-world. This means that an ecolinguistical methodology analyses such stories from the point of view of how stories discursively frame and understand the environment, and how certain discourses and stories suggest certain treatments instead of other treatments of the environment.

The case studies in this book thus employ an ecolinguistical methodology to investigate selected stories that Singaporeans live by. This investigation includes how those stories frame the environment and possibly influence the treatment of the environment of Singapore. This book also pays attention to these Singaporean stories and Singapore's/Singaporeans' treatment of its/their environment as these occur within larger global sustainability trends and discourses about sustainability. Another underlying context of this book, which may not always be foregrounded, is also understanding the stories Singaporeans live by within the rapid unfolding of our collective human-induced climate crisis. This crisis will impact all stories humans live by moving forward. The data is clear that this crisis has been disproportionately caused by certain groups of humans – historically, Global North (also known as Least Affected Peoples and Areas – for more on this, see Chapter 2) countries and especially their wealthy elite; and in this century, by elite and overconsuming countries and communities everywhere. Singapore is firmly in the category of the latter: as shared later in this book, Singapore has a clear class of wealthy elite. Meanwhile, the country as a whole has high rates of per capita greenhouse gas emissions, especially when compared to its neighbours, and most of its economy is based on various processes of consumption.

The overall analysis employed in this book builds upon the three key types of discourses about the environment that a story, discourse, or text may hold. Here, a discourse is assumed to mean 'standardized ways that particular groups in society use language, images, and other forms of representation' (Stibbe 2015, 22). Key to such discourses is recognizing how they encode ideologies about the environment, where ideologies 'are belief systems about how the world was, is, will be or should be, which are shared by members of particular groups in society' (Stibbe 2015, 23). The three possible discourses Stibbe articulates about and that thus represent the environment, and therefore ideologically encode belief systems and ideologies about the environment, are:

1. *Destructive*: those discourses that most strongly ignore the environment and encourage the environment's active (or passive) destruction (these are also the ones that Stibbe says are 'the most important discourses to resist' (Stibbe 2015, 24)).
2. *Ambivalent*: those discourses that attempt to frame or discuss the environment in a positive or healthy way, but that nonetheless still contain or perpetuate an ideology of human exceptionalism; of instrumental views of the environment; and an ideology of capitalist economics that contains within it a story of the environment as only being a resource to be turned via human labour into some type of product or service to benefit the human community; and
3. *Beneficial*: a disruptive discourse that challenges both destructive and ambivalent discourses, and that instead proactively and strategically conveys 'ideologies which can actively encourage people to protect the systems that support life' (Stibbe 2015, 30).

Chapter 3 will undertake an ecolinguistical analysis of various Singaporean texts/sites of stories that are considered *destructive* when compared to the ecosophy this book utilizes. This ecosophy is developed in Chapter 2. Chapter 4 moves into an analysis of various Singaporean texts/sites of stories that are considered *ambivalent* when compared to this book's ecosophy. Lastly, Chapter 5 will do the same, but for Singaporean texts/sites of stories that are considered *beneficial*. Those of us involved in this book project all feel that Stibbe's method of developing an ecosophy and applying it to Singaporean stories allows for a rich investigation of the various ideologies about the environment and society of Singapore that shape how Singaporeans dwell in their ecologies of alive places. It is also a method that allows an ecosophy for the first time, at least according to my knowledge, to be articulated with which to analyse Singapore as a whole. This method of applying an ecolinguistical analysis to stories a country lives by is one all involved with this book hope will inspire readers throughout Southeast Asia (and elsewhere) to do the same for and in their home communities. But we also recognize that this is just one method that could have been chosen as an entryway into an ecolinguistical analysis of Singapore.

Methods of Ecolinguistics

The French poststructuralists Michel Foucault (1926–1984) and Jacques Derrida (1930–2004) were important in developing the methodology of discourse analysis. This is the method this book utilizes, but specifically it undertakes a discourse analysis through the lens of an ecosophy (see Chapter 2). This ecosophy will be systematically applied by the respective authors of Chapters 3–5 to the discourses and texts of Singapore in relation to Singaporean's relations to and understandings of their environment to see if the stories (and their ideologies) are *destructive*, *ambivalent*, or *beneficial*. Those involved in this book are also informed in this choice of method by the religious studies scholar Craig Martin, who explains that discourses carry 'social and political consequences' (Martin 2022, 97). He further explains that,

> discourses are, on the one hand, among the means by which groups wage war; on the other hand, control of authoritative public discourse is a prize awarded to hegemonic groups or institutions. The reproduction of a hegemonic regime requires the reproduction of a discursive apparatus that constructs and naturalizes that very regime. For those who wish to contest such regimes without the use of physical force, alternative discourses must be produced, or the regnant discourse must be appropriated, twisted, and reworked in ways that produce new social objects or entail new relations between existing objects. (Martin 2022, 97)

In the analysis of discourses relating to various environmental places of Singapore, the authors of the case studies to follow are most interested in contesting regimes of anthropocentrism. This is also because anthropocentrism often functions as a form of an interdiscursive, originary and precedential (Silverstein 1996) story by which most people live. This book's various authors also investigate stories about and practices of resource-based overconsumption and how this is a hallmark of ecomodernity, which is another precedential story by which most people live, especially in Singapore. In this investigation to come, we recognize, as do Martin and Stibbe, that *alternative* discourses that challenge both of these must be developed in ways that can create new relations between existing objects (humans and nature) in urban settings during rapid climate changes.

This means this book will critically examine discourses in and from Singapore, especially those framing Singapore as a green exemplar, whether those discourses are from the government itself or from other groups within Singapore, to see if such discourses are *destructive, ambivalent*, or *beneficial*. Given the metrics and issues discussed in Chapter 1 we know that the survival of our species, and the millions of other organisms who also reside on this planet and are 'subjects of a life' (Regan 2004 [1983]), will require us to generate *beneficial* discourses in urban areas to help assist in transitioning to more sustainable dwelling practices in such ecologies of alive places. The critical ecolinguistical analysis of this book is offered in solidarity with this goal, including for the future survivability of Singapore, a country about which we all care deeply. It is also offered in solidarity where our labour is in support of the normative project of shifting academic disciplines (LeVasseur 2021) towards an active support of international climate justice and a just transition (Johnson and Wilkinson 2020). As we respectively write, we thus move between a meta-deictic reference of environmental trajectories that inform our creation of this book-as-text; and micro-deictic reference to Singaporean stories-as-texts informed by the local knowledge (Hanks 2000, 19) of the authors who undertake the various case studies in Chapters 3–5.

It is briefly noted here, too, that one of the key ways discourses shape the *stories-people-live-by* is through use of metaphor, where metaphor often relates to dominator/dominated relationships. Such metaphor even helps to shape the ideology of the audience (Ghdaye et al. 2016). This type of discourse is especially prevalent within a consumer society that has hundreds of millions of dollars spent on corporate advertising and greenwashing (Vieira de Freitas Netto et al. 2020) per annum and that benefits from mass nouns and gendered language that perpetuates anthropocentric and masculine ideologies towards social others and the larger natural world (Stibbe 2004; Fonte 2020; Hultman and Pulé 2019). How nature is discussed and framed by Singaporeans within the context of 'the Singapore Story' (see Chapter 1) through metaphor, but also through homilies, ascription of identity, frames, erasure, salience, and facticity will at times feature in Chapters 3–5. These aspects of grammar and the language that constitute the stories people in Singapore live will be analysed as part of the larger critical ecolinguistical discourse analysis of Singapore's discourses about the natural world undertaken in those same chapters.

In closing this section, I note that the varied approaches of ecolinguistics can be dwarfed by other possible and varied approaches and methods in studying the various interactions between language and culture. As the scholar of cultural linguistics Farzad Sharifian explains, there are three paradigms possible for such study:

1. A focus on documentation and classification of indigenous languages, where language is seen as lexicon and grammar and undertakes studies largely in anthropological linguistics;
2. A focus on how language is used in all available contexts, where language is seen as culturally organized and functions in organizing culture and is studied through linguistic anthropology and the ethnography of speaking;
3. And a focus on identity formation, including narrative and ideology, where language is seen 'as an interactional achievement filled with indexical values' (Sharifian 2019, 4) and is theorized through various social constructivist lenses.[5]

The authors of the case studies have therefore been selective in their ecolinguistical approach. Any project of this book's scope can rapidly become unwieldy and lose its focus. This is because it brings to the study of language and culture, especially paradigms two and three, above, an explicit ecological focus, and within that, a focus on ideology towards the more-than-human world as codified in the stories Singaporeans live by. In this, all involved in this book find Stibbe's approach the best to use, as he is the originator of such an approach that combines deep ecology with an exploration of how the ideology of human domination is grammatically codified in 'texts', broadly conceived. Readers should note that we understand a *text* to be any human-informed creation where values, ideologies and worldviews are embedded in their creation. This means landscapes and ecosystems that humans have managed, created, preserved, built and adapted count as texts, as these places by definition were/are shaped by the ideologies, values and worldviews of the humans whose labour designed and built the place. Within this expansive understanding of 'texts' as discourses to analyse we include national parks, gardens and farms, conservation areas, forests (even the Amazon Rainforest has been selectively co-created by human intention via planting and hunting regimes), to whole bioregional landscapes shaped by human-managed fire regimes. In short, texts are not just written scripts–they are all the things humans create through their labour, wherever that occurs.

Speaking for all the authors, we agree with Stibbe's statement that the environmental[6] crisis 'requires the emergence of a different kind of society, based on different stories. If that is not possible, if the *stories-we-live-by* are too entrenched to be changed, then we will need new stories to survive by, as the earth becomes increasingly hostile to human life and the lives of countless other species' (Stibbe 2015, 183). This normative goal of his method of ecolinguistics – highlighting *destructive* discourses and foregrounding *beneficial* ones instead – is one with which we agree, and that motivated the methodological choices of this book.

With that said, there are other possible methods that can be utilized in undertaking an ecolinguistical analysis of texts and places. For example, it is possible to quantitatively analyse language patterns and lexical repetitions to help understand discourses, including about the environment, as utilized by Richard Alexander in his book *Wording the World*. In this book, he employed a focus on using computer programmes to track and analyse concordancing and corpus linguistic techniques (2008).[7]

It is also possible to undertake comparative perspectives in regards to language–environment interactions. Comparisons can be historical, investigating myths, stories and grammars across time and culture (Verhagen 2008); and they can focus much more on cultures, and the languages and grammars of culture around the natural world. Here, there is often a focus on cultures that have gone through some type of industrial transition, compared to place-based and, often, indigenous cultures that are still largely oral in their language and who actively steward high levels of biodiversity and about which their language is formed (Harmon 1996; Maffi 2005). Another methodology applied in ecolinguistics is to analyse art, poetry, music and other forms of popular culture in order to deconstruct how such forms of popular culture frames and presents the natural world (Stibbe 2012; Ghorbanpour 2016). Similar discourse analyses have been methodologically applied to the education sector, as well, evaluating how published statements and documents discursively frame the environment (Hourdequin and Tani 2017; McNenny 2018) and animals (Jacobs, Teh and Joyce 2016). Another possible method in ecolinguistics includes analysing and tracking language evolution, especially as informed by ecology and the ecologies within which language is learned and structurally changes (Mufwene 2018).

There are also rich possibilities of textual analysis to still come in ecolinguistics, all of which fall outside the purview of this book. These could include analyses of gestures, especially towards non-human others; analysing interspecies play and morality and mourning, including in human-dominated spaces; and how humans will tell the story of possible collapse/s, including intersectional views of such collapse scenarios (Haviland 2001; Knowles 2013; Bekoff 2001; Fix, Burnam and Guttierez 2019; Van Dooren 2016; Kem et al. 2022; Thomas 2022). It is hoped that this book may inform and inspire such granular studies as a next step for ecolinguistics, including such studies undertaken in Singapore, building on this book's varied case studies.

Given the breadth and depth of the above and other possible methodological approaches to investigating language in all its manifestations and contexts, those involved in this project had to pick a methodological approach that allowed us to meet the goals of this edited volume. As ecolinguistics is a new approach to analysing human–nature interactions for the majority of authors, and that all the authors have read and utilized Stibbe's method and approach, it was decided to follow his ecolinguistical method. We all feel that the ability to develop an ecosophy, per Stibbe, both allows for a level of emotional and personal investment into these issues that other ecolinguistical methods do not as easily and readily inspire, as well. Furthermore, this affective impact appealed to our sensibilities as a collective of authors who all have high levels of climate anxiety and who want different *stories to live by*. We also feel that Stibbe's method is easily understood, therefore making ecolinguistics accessible to multiple reading audiences. This, for us, includes the non-academic reading audience/s of Singapore, whom we hope will be able to engage with this book and benefit from its analysis and therefore be inspired by this book to help create a more just, sustainable and resilient Singapore. Lastly, we feel that ecolinguistics is a vibrant and important 'tool' for a multispecies, resilient flourishing toolbox of a shared planetary future. It is also a tool that we encourage more scholars who investigate human–nature interactions, regardless of discipline, to utilize. Therefore, in speaking for everyone involved in this project, we hope our use of Stibbe's method, which we feel is accessible, portable and yet that offers enough rigour and intellectual heft, inspires others to use it as well in undertaking similar analyses of other urban areas. We also hope readers leave this book feeling inspired to critically examine the stories they live by in their own ecologies of alive places and that this inspiration leads to efforts at creating *beneficial* stories in such places. Achieving sustainability in the climate bottlenecks to come will require all communities to undertake such work. This means there must be leaders in those communities doing this work, and we hope readers will readily fill this need.

Overview of Chapters

The ecolinguistical method applied in this book consists of co-authors using an ecolinguistical analysis inspired by the one formulated by Arran Stibbe to investigate actual physical places in Singapore (with some online stories also at times analysed as part of this). Co-authors will use an ecosophy they generated to analyse and evaluate the ideologies and worldviews of those places as codified in the discourses present in the stories being told and therefore lived in these Singaporean places and the texts present in such places. The stories of various Singaporean ecologies of alive places will be judged by the ecosophy to be either *destructive*, *ambivalent*, or *beneficial*. The development of this book's ecosophy (see Chapter 2, that begins with an overview of Western environmental philosophy within which the concept of an ecosophy developed) is central to anchoring such analysis of the ideologies encoded in the various discourses and texts of Singapore. Chapter 3 will analyse *destructive* discourses, Chapter 4 analyses *ambivalent* discourses, and Chapter 5 analyses *beneficial* discourses of Singapore. This book then ends with a conclusion about hoped-for sustainable ways to be humans in urban areas during climate changes. Before developing the ecosophy, however, this book in its introduction first explores why cities are central to human sustainability (or rather, unsustainability) in this century and beyond, and then in Chapter 1 provides a more thorough overview of Singapore and various dominant stories told about Singapore.

Notes

1 Uncle/Auntie is the Singaporean blanket term used to refer to any older person.

2 Or that the cultures within which we live and into which we are socialized, tell us.

3 Here, we include speculative stories about other futures and worldviews, whether in speculative fiction or political treatises.

4 For an overview of the development of ecolinguistics as a field, and how those associated with the field conceive of and operationalize what they do, see LeVasseur (2015), Steffensen (2024a and 2024b) and Zhou (2021).

5 Note that this summary comes from Sharifian (2019, 4, table 1.1).

6 And its interconnected social justice crisis. On how these are linked in Southeast Asia, see Pangsapa (2015).

7 See especially Chapters 1 and 2.

References

Albrecht, Glen. 2007. 'Solastalgia: The Distress Caused by Environmental Change'. *Australas Psychiatry*, 15 Suppl 1: S95–8.

Alexander, Richard. 2008. *Framing Discourse on the Environment*. New York: Routledge.

Bekoff, Marc. 2001. 'Social Play Behaviour. Cooperation, Fairness, Trust, and the Evolution of Morality'. *Journal of Consciousness Studies* 8 (2): 81–90.

Dengler, Corinna, and Lisa Marie Seebacher. 2019. 'What about the Global South? Towards a Feminist Decolonial Degrowth Approach'. *Ecological Economics* 157: 246–52.

Diffey, James, Sacha Wright, Jennifer Olachi Uchendu, Shelot Masithi, Ayomide Olude, Damian Omari Juma, Lekwa Hope Anya et al. 2022. '"Not About Us Without Us"--the Feelings and Hopes of Climate-Concerned Young People Around the World'. *International Review of Psychiatry* 34 (5): 499–509.

Eaton, Heather. 2017. 'The Challenges of Worldview Transformation: "To Rethink and Refeel Our Origins and Destiny."' In *Religion and Ecological Crisis: The 'Lynn White' Thesis at 50*, 121–33. New York: Routledge.

Fix, Adam J., Hugh Burnam and Ray Gutteriez. 2019. 'Toward Interspecies Thinking as a Collabourative Concept: Autoethnographies at the Intersection of Traditional Ecological Knowledge and Animal Studies'. *Humanimalia* 10 (2): 128–49.

Fonte, Diego. 2020. 'Ecolinguistics: The Battlefield for the New Class Struggle'. *Language and Ecology*. https://www.ecolinguistics-association.org/_files/ugd/ae088a_4b5f1f38f397418abf45d70ac4aa8741.pdf.

Ghdaye, Katarina, Valentina Boschian Bailo, Eliana Fortunato, Sidonia Gadient, Ilenia Gosgnach, Mattia Mantellato, Martina Napolitano, Federica Rinaldi and Gemma Roman. 2016. 'Ecolinguistic Perspectives on our Surroundings'. *Language and Ecology*. https://www.ecolinguistics-association.org/_files/ugd/ae088a_7b9a5525016f492b986affc9a8bb4b8b.pdf.

Ghorbanpour, Amir. 2016. 'Ecolyrics in Pop Music: A Review of Two Nature Songs'. *Language and Ecology*. https://www.ecolinguistics-association.org/_files/ugd/ae088a_5e0a2ab7d66049b5b3dd8006e6711225.pdf.

Hanks, William. 2000. *Intertexts: Writings on Language, Utterance, and Context*. New York: Rowman and Littlefield.

Harmon, David. 1996. 'Losing Species, Losing Languages: Connections Between Biological and Linguistic Diversity'. *Southwest Journal of Linguistics* 15 (1–2): 89–108.

Haviland, John. 2001. 'Gesture'. In *Key Terms in Language and Culture*, 83–6. Oxford: Blackwell's.

Hickman, Caroline. 2021. 'Climate Anxiety in Children and Young People and Their Beliefs About Government Responses to Climate Change: A Global Survey'. *Lancet Planet Health* 5: e863–73.

Hourdequin, Peter, and Seiji Tani. 2017. 'The Role of Campus Sustainability Discourse on the Websites of Four Japanese Universities'. *Language and Ecology*. https://www.ecolinguistics-association.org/_files/ugd/ae088a_dfe4b9fdc2704e4680f643ee0deb37ac.pdf.

Hultman, Martin, and Paul Pulé. 2019. *Ecological Masculinities: Theoretical Foundations and Practical Guidance*. New York: Routledge.

Ingold, Tim. 2022. *The Perception of the Environment: Essays on Livelihood, Dwelling, and Skill*. New York: Routledge.

Jacobs, George, Jiexin Teh and Michael Joyce. 2016. 'The Presentation of Animals in English as an Additional Language Coursebooks'. *Language and Ecology*. https://www.ecolinguistics-association.org/_files/ugd/ae088a_43d39bc18b0b43f9ac1ab11ac4af8339.pdf.

Johnson, Ayana, and Katharine Wilkinson, eds. 2020. *All We Can Save: Truth, Courage, and the Solutions for the Climate Crisis*. New York: One World.

Kemp, Luke, Chi Xu, Joanna Depledge, Kristie L. Ebi, Goodwin Gibbins, Timothy A. Kohler, Johan Rockström, et al. 2022. 'Climate Endgame: Exploring Catastrophic Climate Change Scenarios'. *Proceedings of the National Academy of Sciences* 119 (34): e2108146119.

Knowles, Ric. 2013. 'Editorial Comment: Interspecies Performance'. *Theatre Journal* 65 (3): i–v.

LeVasseur, Todd. 2015. 'Defining 'Ecolinguistics?': Challenging Emic Issues in an Evolving Environmental Discipline'. *Journal of Environmental Studies and Sciences* 5: 21–8.

LeVasseur, Todd. 2021. *Climate Change, Religion, and Our Bodily Future*. New York: Lexington Books.

Maffi, Luisa. 2005. 'Linguistic, Cultural, and Biological Diversity'. *Annual Review of Anthropology* 34 (1): 599–617.

Martin, Craig. 2022. *Discourse and Ideology: A Critique of the Study of Culture*. New York: Bloomsbury Academic.

McNenny, Gerri. 2018. 'Framing the Anthropocene: Educating for Sustainability'. *Language and Ecology*. https://www.ecolinguistics-association.org/_files/ugd/ae088a_d0a2988d98dc4a52807470e41726bab8.pdf.

Mufwene, Salikoko. 2018. 'Language Evolution from an Ecological Perspective'. In *The Routledge Handbook of Ecolinguistics*, 73–88. New York: Routledge.

Oelschlaeger, Max. 1994. *Caring for Creation: An Ecumenical Approach to the Environmental Crisis*. New Haven: Yale University Press.

Pangsapa, Piya. 2016. 'Environmental Justice and Civil Society: Case Studies from Southeast Asia'. In *Routledge Handbook of Environment and Society in Asia*, edited by Paul Harris and Graeme Lang, 36–52. New York: Routledge.

Ray, Sarah Jaquette. 2020. *A Field Guide to Climate Anxiety: How to Keep Your Cool on a Warming Planet*. Berkeley: University of California Press.

Regan, Tom. 1983. *The Case for Animal Rights*. Berkeley: University of California Press, 2014.

Sharifian, Farzad. 2019. 'Language and Culture: Overview'. In *The Routledge Handbook of Language and Culture*, 3–17. New York: Routledge.

Silverstein, Michael. 1996. 'The Secret Life of Texts'. In *Natural Histories of Discourse*, 81–105. Chicago: University of Chicago Press.

Steffensen, Sune Vork. 2024a. 'On the Demarcation of Ecolinguistics'. *Journal of World Languages* 10 (3): 499–527.

Steffensen, Sune Vork. 2024b. 'Surveying Ecolinguistics'. *Journal of World Languages*.

Stibbe, Arran. 2004. 'Masculinity, Health and Ecological Destruction'. *Language and Ecology*. https://www.ecolinguistics-association.org/_files/ugd/ae088a_e2e39691a7874fc3b46b99d24edffd5d.pdf

Stibbe, Arran. 2012. 'Chance Encounters: Ecology and Haiku-Inspired Photography'. Working paper. https://www.ecolinguistics-association.org/_files/ugd/ae088a_2dd2ecbf8e54492190e36153e462d1b8.pdf.

Stibbe, Arran. 2015. *Ecolinguistics: Language, Ecology, and the Stories We Live By*. London: Routledge.

Thomas, Leah. 2022. *The Intersectional Environmentalist: How to Dismantle Systems of Oppression to Protect People + Planet*. Boston: Little, Brown.

Van Dooren, Thom. 2016. *Flightways: Life and Loss at the Edge of Extinction*. New York: Columbia University Press.

Verhagen, Frans. 2008. 'Worldviews and Metaphors in the Human-Nature Relationship: An Ecolinguistic Exploration through the Ages'. *Language and Ecology* 2 (3). https://www.ecolinguistics-association.org/_files/ugd/ae088a_7d7264a0c6ea4e3fbcba1270a50260c0.pdf.

Vieira de Freitas Netto, Sebastião Vieira, Marcos Felipe Falcão Sobral, Ana Regina Bezerra Ribeiro and Gleibson Robert da Luz Soares. 2020. 'Concepts and Forms of Greenwashing: A Systematic Review'. *Environmental Sciences Europe* 32 (1): 19.

Zhou, Wenjuan. 2021. 'Ecolinguistics: A Half-century Overview'. *Journal of World Languages* 7 (3): 461–86.

Acknowledgements

Land Acknowledgement

The authors of *Critical Ecolinguistics Within Singapore, 'the Garden City'* wish to humbly acknowledge the land we have resided on – a land imbued with memory, resilience and life. We are deeply privileged to honour its spaces through the stories we tell. As writers, we bring diverse connections to Singapore. Some are descendants of immigrants who journeyed challenging paths to settle on this land. Others arrived on Singapore's shores within their own lifetimes. What unites us is a shared gratitude and deep indebtedness to this land – for providing shelter, care, community and a place to call home for ourselves and our families.

We recognize that the transformation of Singapore into a bustling metropolis today was forged through violence towards its natural environment and the Indigenous communities that inhabit it. Forcefully embracing modernity and urban redevelopment has continually displaced the physical homes and traditional ways of life of communities like the Orang Laut and Orang Pulau. Let us honour their wisdom, resilience and traditions, whose knowledge and relationship with the land have left an indelible mark on this nation's history. We express our gratitude for the foundations they have laid and for their stories, both remembered and untold, that continue to guide and inspire us today.

We also acknowledge the more-than-human, who continue to persist in the face of rampant ecological destruction. We coexist with Singapore's forests, mangroves, wetlands, waters and streets with our friends – the Sunda pangolin, the smooth-coated otter, the paradise tree snake, the mudskipper, the moon crab, the olive-backed sunbird, the Asian toad, the nudibranchs and the neighborhood cats. They are the beings that share the blue, the green, the grounds beneath us and the skies above us.

We give thanks to the land for imbuing us all with life despite how Mother Earth increasingly bleeds and suffers because of our collective mistakes. We are forever grateful for the opportunities the privilege of being on this land has afforded us and vow to continually give back to these spaces and their inhabitants.

Individual Acknowledgements

Anjali would like to thank the other authors of this book for their support and unwavering belief in it. She is grateful for the community that acted as a catalyst for the creation of this book – at Yale-NUS College – where she learned that constructive criticism should come from a place of love. She would like to thank her friends and family for their constant encouragement and guidance.

Beverley is grateful to Yale-NUS College, a place that has forever transformed the way she views and interacts with the world. They would also like to thank the other authors of this book (what a journey this has been!) and Bianca and Catherine for sharing their insights about Bukit Brown with her. Finally, their friends and family, the ocean, forests and critters that continue to inspire awe and love in her, fueling her work to fight for climate justice.

Hannah thanks every human and non-human who has extended encouragement, inspiration and critical thought; especially the Yale-NUS College community, the coral reefs and intertidals, the activists and advocates and the environmental and biodiversity communities in Singapore, especially those who took special time to share their knowledge with her. I stand on the shoulders of giants. She thanks her family and partner for putting up with her through a crazy year of writing, and Singapore, the place she has always called home. And of course, *Ad Majorem Dei Gloriam*.

Kristina thanks her community and ecosystem for the making of this book. Writing has always been her labour of love, and this writing in particular feels like it comes from a shared love that spurns our collective struggle for something better. Thank you to the co-authors for writing this love letter to Singapore (of sorts) together. To all the community organizers and activists in Southeast Asia, the land we live on and with and the oceans and hopes that connect us all – thank you, and here's to writing and fighting alongside each other, for liberation.

Lecarl thanks the Yale-NUS College community for its transformative role in his learning and the invaluable support and guidance he received throughout his journey. He extends a special thanks to the friends who inspired him and stood by him in the Inter-University Environmental Coalition and Post-Growth Singapore, as well as other environmental advocates and community organizers. Their energy, knowledge and friendship have invigorated and empowered his work.

Yock extends his heartfelt gratitude to the YNC community for its time, space and unwavering support throughout his four years within its corridors. Deepest appreciation goes to those at home, whose constant encouragement has guided him on this uncertain yet meaningful path of environmentalism. Finally, he lovingly remembers his dogs, Tyson and Tigger, and honours those who have passed – Lulu, Nemo, Niu Niu and more, who have taught him loyally the true meaning of unconditional love and how to cherish the non-human who share our lives.

Todd wishes to thank the Yale-NUS community – its students, staff, faculty and upper-level administration (especially: David Jacks, David Post and President Joanne Roberts) – for their support, hospitality and in making a world-class learning community. Thanks also go to the residents and staff of Kent Vale, as well as to Ain and Louis, the 'KV Ballers', the Elm lunch crew and the Environmental Studies team (esp. Michael Maniates and Marvin Montefrio). An extra special thanks for their friendship goes to Kyle Gibson, Max Boykoff, Jonathan Vandenberg, Ryan Tans and Yong Zhang. Final thanks goes to Christoph Draeger for lots of talks all over Singapore about music, collapse, art and empire. At Bloomsbury, he offers immense gratitude to editors Sarah MacDonald and Laura Gallon for their patience, many emails and stewardship of this project; and to M. Agalya, for their assistance with copyediting; and to Arran Stibbe, for believing in this project, his many years of friendship and the example of his deep love of the earth and how this love can be centred in academic labour.

Introduction: Cities in and of the Climate-Changed Future

Todd LeVasseur and Hannah Ng Shueh Yi

Perspectives on Climate Change in Cities

All humans reside in plural communities where people live by different stories and ideologies. Therefore, there exists different perspectives about and internal discourses towards all the events people experience, with this difference based in large part on different personal and group ideologies. Such differences also affect how different communities experience and understand anthropogenic climate change. This applies regardless of whether certain communities accept climate change as a human-caused biogeochemical phenomenon and thus of the lived effects of climate change on one's life, or if the community does not. Regardless, each person's perspective varies to some degree. This is especially true based on geographic location. For example, an Indonesian might accept human-caused global warming as a serious phenomenon; not only is their country a nation of archipelagic islands, but their capital city, Jakarta, is at high risk of flooding due to climate change-induced sea-level rise and overexploitation of their water table. This has led to the ongoing relocation of the Indonesian capital city to Nusantara, on Kalimantan, another island. On the other hand, someone from the USA exposed to Republican-supported intransigence and disinformation campaigns related to anthropogenic climate change may be among the significant number of Republican voters who deny humans are impacting and changing the climate (LeVasseur 2012; LeVasseur and Zaleha 2019). Given the focus of this book, in this chapter, we explore a few climate stories and perspectives that are currently present in Singapore.

As you have chosen to read this book, we assume that you have some understanding of the severe shocks our ecosystems face due to anthropogenic climate change. If so, then many narratives in this chapter may be familiar. Despite this familiarity, we invite you nonetheless to contemplate the narratives about climate change presented in this chapter. In doing so, we hope you can discover how and why analysing climate change through ecolinguistics might change your perspectives on it, and more broadly, your perspectives on the world around you. This chapter is also of importance as any sustainable future of Singapore (or any global city) will require the country to both mitigate climate impacts and adapt to climate destabilization. Therefore, how people interpret and respond to climate change narratives is of importance, for two key reasons. The first is because the process of interpretation and response has important ideological impacts, including on the perceived severity of climate futures. The second is how people interpret and respond to climate narratives will have significant political impacts (LeVasseur and Kingston 2024).

The Climate Science Narrative

Drastic planetary climate changes have occurred throughout geological time. Periods of glacial cooling, then warming, have repeated in cycles and have moderated Planet Earth for aeons. However, in the short two-and-a-half centuries since the Industrial Revolution, the planet has entered a period of rapid and drastic overall warming. Through referencing paleoclimates and modelling current trajectories, climate scientists have found that the current climatic conditions are warming at an anomalous rate compared to any time period in the last one thousand years (Mann, Bradley, and Hughes 1999). This period of warming on Earth will result in extreme climatic changes. This includes a prediction of much more extreme weather patterns that will likely cause the disruption of life as we know it. Science has proven that, without a doubt, the Earth and almost all of its inhabitants are in dire straits given such amplifying climate trends. By the 1970s–1990s, climate scientists, biologists, naturalists, and others knowledgeable in ecology began to see warning signs of a changing environment. In 1990, the United Nations Intergovernmental Panel on Climate Change (IPCC) released its first global climate assessment (IPCC and WMO 1992). This laid important groundwork for later international agreements on climate issues but also for the development of other environmental and theoretical frameworks to understand climate change.

One such framework was the Planetary Boundaries – first published in 2009, this framework identified nine critical processes and associated thresholds for the Earth to maintain its stability and resiliency (Rockström et al. 2009). In

2023, the biological oceanographer Katherine Richardson and others from the Stockholm Resilience Centre found that Earth, mainly because of anthropogenic activities, had already crossed six of these nine boundaries (Richardson et al. 2023) These boundaries sadly do not yet account for species extinction rates, which are at sobering levels and suggest an evolving world that will be much less biodiverse. For professionals in climate science, policy, and advocacy spaces, these are flashing danger signals. The boundaries schema can help to turn people's attention to understanding the global, interconnected scale of climate change while also highlighting urgent areas of needed intervention.

These two science-backed narratives of climate change centre our understanding of environmental issues around science-based studies, built upon observations and climate models made by the scientists who study and measure Earth's biogeochemical cycles. This particular story about anthropogenic climate change is thus science/evidence/fact-based and tells the story: Humans are causing the climate to change while humans are also overshooting safe planetary boundaries and processes.

The Lived Narratives of Climate Change

Other perspectives and lived experiences of climate change create different narratives. For example, one of the top Singaporean comments is 'Wah, today damn hot'. To put this into numbers, the global average temperature in 2023 was 14.98°C (ECMWF 2024) as compared to Singapore's 28.2°C. Singapore's temperatures have trained me (Hannah) to be able to tell between a 31°C to 32°C daytime temperature – above 31°C, and my own body begins to feel the strain of heat. A dominant perspective and narrative on climate change in Singapore is therefore deeply entangled with the physical discomforts of severe to extreme heat, to say nothing of increasingly erratic rainfall events. Science-based narratives do not capture the lived reality of a changing climate that Singaporeans experience: more heat and humidity do not show up on an abstracted X–Y axis. Our guess is this holds true for many readers of this book who are having their own new experiences with a changing climate – this is probably especially true for readers aged forty and over. Heat vulnerability can also take on violent effects, as it is differentially experienced in Singapore according to age, class, and labour type (Berastain 2024). For example, heat-vulnerable populations include the elderly population in Singapore (of which there is a growing population in Singapore's ageing population demographic) who are subject to more heat sensitivity, as well as outdoor construction workers who are often foreign migrant workers and who are exposed to extreme levels of heat in their day-to-day labour.

In Singapore, a country with a largely scientifically literate citizenry cultivated through a country-wide education system that includes climate change in its syllabi from ages seven and up (Environmental Education 2019), most people understand climate change in ways informed by climate science (see above). Such climate literacy is coupled with the lived experiences of Singaporean residents, where Singapore is an equatorial country that experiences heat vulnerability. Its geography means that for many in Singapore, there are particular and immediate narrative perspectives brought about by such geographical and physical realities that impact the felt, lived experience and thus understanding of climate change in ways that more 'dry' science-based studies often do not. For example, according to the Meteorological Service of Singapore, 2023 was our fourth warmest year on record since 1929, and the third consecutive year of breaking the decadal mean temperature, with five of the top warmest years for Singapore occurring in the past decade (Meteorological Service Singapore 2024). Singapore maintained a toasty annual daily maximum temperature of 31.9°C in 2023, and considering Singapore's humid climate, this makes for a rather swampy, sweaty city (and exacerbates Singapore's greenhouse gas emissions as these temperatures drive demand for air conditioning and cooling technologies).

On scales of heat alone, Asia is warming faster than the global average. In April 2024, heatwaves hit South and Southeast Asia at record-breaking temperatures. These trends continued apace from 2023, with many countries in the region recording months of record-breaking heat. Impacts of such record heat included power outages, school shutdowns, and other infrastructural stress that further impacted those already without the resources to keep cool and those whose livelihoods rely on natural resources (Chen 2024; Magramo and Chen 2024). Populations dependent on agriculture, many of which are in Southeast Asia, struggled as their fields and rivers dried up. The years 2023 and 2024 have followed the same rising trends in temperature since the 1990s, with no expectations of cooling seen on the horizon. Therefore, almost all human lives and livelihoods are continually and increasingly impacted throughout Asia by these patterns, so that climate change is no longer a faraway concept; it is already impacting lives, now, and this will unfortunately continue in the decades to come. The lived heat experiences in Asia show us that climate change is not just some faraway impending doom but rather a crisis that is killing people in the present and in unevenly distributed ways.

Unsurprisingly, this has added a level of climate anxiety to many of us in Singapore who are aged under thirty. Such climate anxiety therefore is now increasingly becoming part of any climate narrative or climate experience, as well (Diffey et al. 2022). Such anxiety is also seen with climate scientists too. For example, in a May 2024 survey of 380 climate experts, only twenty-three believed we would be able to limit overall planetary warming to 1.5°C. With temperatures projected to continue to increase, the climate crisis will cause more deaths and affect more people's health than the damage and harm that has already occurred, especially in our part of the world.

The Common but Differentiated Responsibilities of Climate Change

The year 2024 was the warmest year on record, according to the EU's Copernicus Climate Change Service (C3S). The year 2023 was (also!) the warmest year on record by an additional 1.45 ± 0.12°C across the globe (WMO 2024). This is dangerously close to the 1.5°C goal of the Paris Agreement, which is the overall amount of warming deemed acceptable by 2100 – a goal for seventy-five years in the future (at the time of writing) (UNFCCC 2015).[1] The 1.5°C goal is a science-based international goal that is held together by negotiations at the Conference of Parties to the United Nations Frameworks for Climate Change (UNFCC) – a non-binding political process of change by which countries, acting as parties, can incentivize reduction of emissions among other climate adaptation and mitigation strategies as per the Paris Agreement to limit anthropogenic impact on the planet.

However, a serious drawback of climate change impacts, especially when it comes to mobilizing political responses, is that they are largely unevenly distributed. Climate impacts differ based on regional geography, abiotic features of the land, and the ability of communities to adapt to these effects. And, of course, as will be explored more thoroughly below, given colonial and imperial histories, there are extremely uneven causes behind which human communities and activities have historically created climate change from the start of the Industrial Revolution to today. The key here is to note that centres affected by climate change today are vastly different from the centres of current and historical emissions.[2] Temporality complicates this further: emissions from the industrial age persist, whereas younger generations bear the brunt of its effects; and yet unborn generations of all species, including humans, will bear even more of its effects. Therefore, geographies that have suffered under historical oppressions tethered to colonial rule and have had less time to 'develop' are tragically and devastatingly unprepared for their populations to adapt to emerging and future climate change impacts. However, the physical and geographical realities of colonized places, especially in the tropics like Singapore, often put these communities on the front lines of dealing with climate change realities.[3]

Among those frontline communities are many within Asian countries – the IPCC found that Asian urban areas are considered 'high risk locations from projected climate change, extreme events, unplanned urbanisation and rapid land-use change…By 2080, 940 million to 1.1 billion urban dwellers in South and Southeast Asia could be affected by extreme heat lasting more than 30 d yr −1, with poorer populations affected the most' (IPCC 6AR 2023, 1460[4]). Moreso, one in three climate migrants already come from Asia, and 'the highest ratio of outward migrants is seen from hazard-exposed Pacific countries' (IPCC 6AR 2023, 1469).

This particular geographical spread has led to diverging priorities: when it comes to responding to anthropogenic climate change, people and their governments cannot agree on what to do or how to do it. Such disparate responses are formed by unique regional perspectives and narratives regarding the ability of populations to adapt, as well as by economic and political constraints and possibilities. In Singapore, technocracy and ecomodernism hold strong, with the National Climate Change Secretariat identifying areas of 'Green Growth Opportunities' including 'Clean Technology' as important strategic investments for Singapore (NCCS 2024).[5] For example, in 2024, Singapore's Deputy Prime Minister Lawrence Wong (who at the time of writing was the acting Prime Minister of Singapore) stated, 'We are techno-optimists by nature' (Garekar 2023), portraying Singapore's ecomodernist worldview.

On the other hand, developing countries emphasize the Common But Differentiated Responsibilities (CBDR) of the 2015 Paris Agreement as they pursue rapid development pressured by growing urban populations facing vulnerabilities from climate change.[6] These negotiations, which happen at the country level at conferences related to the Paris Agreement and other United Nations climate-related conferences, end up being dominated by the most stubborn actors who are the least impacted and who have the most to lose in agreeing to carbon emissions reductions (Falkner 2016). Despite their success with the chlorofluorocarbon reductions that saved the ozone layer from destruction in the 1987 Montreal Protocol on Substances that Deplete the Ozone Layer global treaty, these international platforms have not been able to address the climate crisis and the urgent situation we now face – with scholars analysing this inability as a collective action/free-rider problem (Roy 2021).

With this brief overview of anthropogenic climate change and some of its possible impacts on Singapore and its geographical region in place (note: this theme and this data run throughout the entire book), we now move into a brief exploration of the dominant narrative perspective on climate change that we face primarily in Singapore. Given the lived experiences of this book's authors (especially those from Singapore), this hegemonic narrative also informs the ecolinguistical analyses in Chapters 3–5.

The Human-Centric (Consumerist, Urban) Perspective

According to the United Nations, urban populations have grown from 751 million in 1950 to 4.2 billion in 2018 (United Nations 2018). This burgeoning growth in urban populations has resulted in metropolitan lifestyles/discourses based on high levels of consumption proliferating through these urban populations. Urban settlements are usually defined by a high population density within an area, often clustered in cities. Cities constitute a starkly different lived reality than rural environments, where living in a city influences, changes, and shifts a person's interaction with the natural world in ways very different than how one experiences the natural world in rural environments.

Cities inherently require massive amounts of material and dense energy to maintain their baseload needs: energy, housing, food, and transportation.[7] While perhaps appearing to condense the needs of populations into a smaller physical footprint, their dependence on mass production and supply chains often hamper any forms of achieving true sustainability (defined and explored below). Instead, linear fossil-fuel-based industrial production (both in the Marxist and supply chain senses) caters to mass consumption while density is privileged over intentional urban planning (the latter is one of the 'success stories' of Singapore, at least as marketed to international audiences). This is the only way to date to feed consumer-based urban lifestyles. Taken together, industrial modernity's reliance on technology depends on high rates of unsustainable consumption and the rapid disposal of natural resources. These combine so that products sold and used in urban areas come from unsustainable processes of production that are stretched across massive supply chains (or as economists may call them, global value chains; on this, see Seric and Yee 2019). Too often, the high-emitting corporations that have created these global urban supply chains externalize the 'true costs' of these products. This means that many global corporations elude responsibility for both the social and environmental harms and externalities that the industrial processes used to make their products cause.[8] The result is unsustainability, so that even the most 'green' and 'sustainable' products touted by a large company are due a healthy suspicion from a strong sustainability perspective. Cities, now and into the future, will have to deal with these social and environmental externalities.

Cities of the Future

Overall, the trajectories related to human-induced climate heating are a threat multiplier that will impact all habitats for all extant species the world over (Fletcher et al. 2024; Merz et al. 2023). Given demographics and migration patterns, these impacts are those that cities and island-nations will have to either mitigate or adapt to the world over, including in Singapore (He, Zhang and Yuan 2022; Aik et al. 2023; Bhullar 2013). According to Professor Winston Chow, the first Singaporean nominated to IPCC leadership, the predicted impacts of climate change on cities suggest that 'Critical infrastructure within settlements such as transportation, water, sanitation, and energy systems [will be] compromised by extreme weather events. Cities and settlements by the sea are specifically impacted by climate hazards. They are at the front line of climate change, being directly exposed to interacting climate and non-climate coastal hazards, such as sea-level rise and destruction of local ecosystems that previously protected people living along the coast' (Ipsun Solar 2022).

These are also the trends that the various authors in this book, who other than Todd are collectively part of the climate generation (Ray 2020), have known their whole lives. Living with the spectre of climate destabilization in an industrialized, highly urban country for these authors directly informs parts of the ecosophy shared in Chapter 2 and also directly informs felt, lived experiences of climate trauma and grief in regards to possible climate triage facing Singapore's future by 2100. As shared above, trends suggest over 80 per cent of humans by the end of the century will live in urban areas, leading the Organisation for Economic Co-operation and Development (OECD) to call this 'The Metropolitan Century' (2015). In many ways, then, how Singapore, assumed to be a 'green urban exemplar' (see next chapter) experiences climate change will be a harbinger for how many other cities will experience climate change. However, many of these cites, especially in Southeast Asia, do not have the resources or technology that Singapore currently has at its disposal to navigate climate change.

The human future is also going to be shaped by supply chain disruptions, resource limitations or new inventions, and climate-influenced resource shortages (especially related to water), and it will unfold within the context of continued biodiversity collapse (of broad taxa, including whole aquatic and terrestrial ecosystems, especially in the cryosphere). Humans are going to be required to navigate this biogeochemically and economically constrained future in some capacity. How this occurs will impact any type of resilient, regenerative, sustainable, healthy, and just society that emerges. Moreso, these interrelated factors related to any type of possible flourishing in the climate-changed future will be determined in large part by metrics applied to urban living. This holds especially for Singapore and its neighbours in Southeast Asia.

Just like our readers, we as authors of this book have been living within various narratives around climate change and climate action. Some of these have been briefly explored here. And just like some of our readers, many of us have faced existential dread seeing our planet teetering on the edge of disaster – and yet we have known no other world. It is with this in mind that we have embarked on our own ways to be the change that we want to see, leading us respectively to matriculate at Yale-NUS College and to embark on the journey of writing this book. For the authors, this process of responding to climate change included receiving degrees in environmental studies. It has also spilled beyond education into hobbies, lifestyles, starting climate justice organizations, and many other choices informed by climate science and the lived experience of climate change in Singapore. It is from this unique position of our collective stories and climate narratives that we contribute through this book to a critical understanding of the currently unsustainable world around us, but a critical understanding grounded in Singapore. In doing so, we shed light on situations and stories that may seem black and white, but upon closer look, unveil a spectrum of grey. We also do this while centering and honouring the work of many others, especially those in Most Affected Communities and Areas, and those who have done and continue to generate decolonial, climate-adaptive lifeways. Ecolinguistics aids in these normative goals, and this insight inspires our storytelling in this edited book.

Why Singapore?

This question is, of course, explored throughout the entirety of this book. Here, it is important to note that Singapore's unique sociopolitical system has been instrumental in driving its many 'green' initiatives, narratives, and ideologies that impact all of its citizens. Nevertheless, what exactly does celebrating Singapore (or any other leading city in sustainability) really mean in the context of this century's rapid climate changes?

To answer this, we must first operationally define sustainability. To us, sustainability is the capacity and ability of human communities to design their lives such that the social, environmental, and economic needs of the human community are met in perpetuity. Additionally, in so doing, society must not cause or further social disparities and injustices, all while allowing for jobs and industries that pay a living wage, thus contributing to economic (and thus social) flourishing in ways that do not compromise ecosystem integrity.

This is an aspirational goal building on the concept of sustainability as the triple bottom line (Elkington 2004) of social, environmental, and economic systems and their interaction. It is also an understanding of sustainability that navigates between a Brundtland paradigm of sustainable development and a resilient paradigm of sustainability (Thompson 2016). While the ecosophy generated in Chapter 2 skews more towards the latter paradigm of resilience and strong sustainability (see below), the reality is that we understand that cities are currently still driven by concerns of economic growth and hence advocate weak sustainability (see below).

This is partly because being elected, appointed to, and maintaining positions of power within governmental apparati is difficult while campaigning against an economy-first paradigm of weak sustainability.[9] Most people all over the world still live under the paradigms of economic growth – prosperity, and prosperity – the ability to meet the needs of the electorate. This, coupled with logic and pressures of international competition, compels most extant governments to ensure that their workforce and, if applicable, their country in general stays fiscally competitive.[10] Regardless, all cities engage in the practice of cultivating imaginaries (*stories-we-live-by*) in their citizens, ranging from cornucopian economic growth, to those of environmental inattention and inaction to make invisible and thus hide the destruction of the natural world that city/urban living requires (Carr and Milstein 2021).

An exploration of climate-changed future planetary realities for urban areas within the context of sustainability is important for other interconnected reasons: (1) this book is a sustained ecolinguistical analysis of a city, and to our knowledge, this is the first such analysis of its kind; (2) to be thorough in this task, we must attend to one of the key global dynamics that will determine not only Singapore's environmental and biosocial future, but the future of any and every urban area the world over: we must attend to anthropogenic climate change and its current and future impacts on urban dwelling practices (Ingold 2000). To do this means investigating past and present causes of anthropogenic climate change and the roadblocks to stopping climate change; (3) Singapore does not exist in isolation but rather its location is highly vulnerable both geopolitically, but also ecologically, where sea-level rise, water shortages, changing rain patterns, an increasing heat index, and changing migration patterns of non-human earth residents will all impact Singapore's future; (4) Singapore's government has an existing Climate Action Plan (2016) that tells its own story about Singapore's plans to adapt to climate change – a thoroughly ecomodern story that is laudable, but that also often falls short based on our ecosophy; (5) and lastly, we must be honest about the sobering bottlenecks most likely coming our way as a species due to climate change.

On point five, Singapore's precarious position at the tip of Southeast Asia, its current strategic import as both a key node of petroculture (Wilson, Szeman and Carlson 2017) and in the enactment of petromasculinities (Daggett 2018), and being a key centre for global capitalism, all will influence Singapore's efforts to mitigate and adapt to climate change. Furthermore, the sobering likelihood of increased climate refugees in the region (Islam and Khan 2019; Elmhirst, Middleton and Resurreccion 2017) and the politics of such refugee flux (Marthin and Budiman 2020) are also key climate-induced bottlenecks that will likely determine key parts of Singapore's biosocial future. Taken together, these five points require us to dig deeper into these same dynamics throughout this chapter, to help situate the larger goals and trajectories of this book, as well as some of the key goals this book intends for the series of which it is part. In short, every city of the future, and therefore, the island city-nation-state of Singapore, will be a climate-changed city. Understanding this is key in helping situate our ecosophically informed ecolinguistical analyses that follow in subsequent chapters.

Climate-changed cities, especially with largely urban, globalized populations, and their efforts towards shifting to sustainability, deserve a more sustained exploration. We hope our book helps contribute to this larger needed exploration (Kemp et al. 2022; Heikkurinen, LeVasseur and Ruuska 2022; Brozovic 2023) and can highlight how ecolinguistics can assist such analyses and efforts towards strong sustainability (Bonnedahl and Heikkurinen 2018).

The Need for Stories of Resilience

A key insight of ecolinguistics is that the *stories-we-live-by* are never neutral. As they contain ideologies, they influence how we see the world, act in the world, and, thus, treat the world. They influence how we treat other beings, organisms, and landscapes with whom we individually and collectively share the world. Therefore, the stories Singaporeans live by, especially the stories of and about Singapore told by its government and other peoples, matter. They matter especially to the approximately six million people living in Singapore (citizens, foreign workers, and tourists, all); the abiotic and biotic places and lifeforms that make up Singapore; and to the future Singaporeans who will have to navigate climate change and other sobering environmental metrics as urban peoples of the future.

Singapore, like many other urban areas, has a dominant ecomodernist story in visions of mitigation and adaptation. Ecomodernization believes it is possible to have a 'good Anthropocene' (Asafu-Adjaye et al. 2015) and believes that humans and human development do not inevitably lead to destruction and environmental collapse. Rather, ecomodernization maintains that humans are (or at least, possibly can be) healthy to the planet while generating flourishing human lives. This optimistic reading claims that continued human ingenuity, ethics, planning and design, smart policy and regulation, sound incentives, and strategic use of science and technology can and will lead to better social health, economic growth, and environmental protection.

Rather than a 'doom and gloom' vision of Malthusian contraction, ecomodernists claim that the Kuznet curve[11] holds constant for most new technologies; over time, their environmental metrics have and will improve. Their counterpoint to 'doom and gloom' though is that with more ingenuity, efficiencies, access to markets and incentives for sustainable design and technology, coupled with sound environmental policies and liberal democratic managerialism, improvements will continue to be made and new technologies yet to be invented will assist even more in decoupling human well-being from environmental harm. Ecomodernists point to successes in forestry, agriculture, energy production, public health, and education in support of their analysis; for the latter three, Singapore is often a poster child. Ecomodernists are not naive in their analysis and recognize improvements must be made in all of these areas but hold to the core belief that 'both human prosperity and an ecologically vibrant planet are not only possible but inseparable', especially as human well-being is decoupled from environmental destruction.[12]

Apart from ecomodernism, there are many other ideologies, narratives, and *stories-we-live-by* that deserve attention that will be encountered in the analyses undertaken in this book. Importantly, apart from the ones presented here, there is still a need for stronger, new, *beneficial* postcolonial and decolonial urban stories in an era of rapid climate changes. These stories tie into larger tensions that are briefly raised in this chapter, but that have implications for the entire book. These are tensions between the stories humans live by now and those of the near future, especially in urban areas where future stories will be impacted by the ever-increasing traumas brought by climate destabilization. A particularly deep tension motivating this book is how humans, especially in Singapore, will move into such challenges in the future. Will navigating these climate challenges be done through technology, bureaucracy, optimistic use of ingenuity, innovative engineering, and an abiding faith in the economy? Or will such navigating be done with a continued and deepening critique of capitalist growth (especially neoliberal forms of capitalism) and accepting that there are resource limits that as a species we must live within, as contained in a degrowth perspective? In other words, will it be an ecomodern future of decoupling human ingenuity and quality of life from environmental destruction via technological innovation and human ingenuity (such as the future envisioned by Singapore) as briefly articulated above? Or will it be one of multispecies and new materialist insights of entangled co-becoming (LeVasseur 2021), where conceptions of a good life are entirely uncoupled from growthism and where the state provides public goods for all, rather than providing incentives for unlimited consumption and competition?

Another key point in this line of questioning is that as biophysical ecologies of alive places are materially transformed, the 'metabolic action of human labour in transforming material nature thus also involves the production of ideas, representations, and understandings of nature, featuring a cultural politics of meaning that is internal to the social production of nature' (Ekers and Prudham 2017, 1379). This implies that ideologies and the stories we live by dialectically inform and are informed by how humans produce their habitats and impact the environments around them. This book is in solidarity with other ecolinguistical analyses in calling for beneficial *stories-to-live-by*: We need to represent and produce ideas and ideologies of the more-than-human world that encourage strong sustainability.

While we applaud the hopeful storytelling of ecomodernization, we recognize that it does not take seriously enough both the metabolic rift and green colonialism (Singh 2021; Zografos 2022) and thus does not grapple seriously enough with the biogeochemical realities of rapid and alarming ecosystem collapse. Moreover, within now-global patterns of overconsumption and subsequent Earthen destruction, the story of ecomodernization too often ignores the reality of 'Jevon's paradox'. The paradox itself is that efficiency gains based on technological advances at any step of a supply chain (or in the energy sector broadly) not only do not reduce the demand for and consumption of products but rather paradoxically almost always increase such demand and consumption, especially at scale.

This suggests that more efficient technologies and ongoing hoped-for 'sustainable' consumption will not help humans live within the bioregenerative carrying capacity of the planet, nor will it reduce overall emissions to meet the targets of staying under 1.5°C global temperature rise. The tensions inherent in Jevon's paradox are further amplified in a key ideology of global society where the expectation of consumption is pegged to global social progress and development. This story of ecomodern progress and consumption as being evidence of a good life is very evident in Singapore and is a story that the co-authors of this book have been raised within as a key ideology/*story-to-live-by*. The result is that we, like most communities everywhere, are trapped in an endless cycle of overexploitation and consumption, where the result is that we are hurtling faster towards massive climatic disruption/(s). Rather, our various ecolinguistical analyses of Singapore recognise that there are very real planetary boundaries we cannot innovate our ways out of and that 'The ideologies of colonial-capitalist renditions undergirding economic growth models [including ecomodern ones] are incommensurable with sustaining any planetary boundaries' (Sultana 2023, 4).

In closing this chapter, we note that it is important to analyse the feasibility of Singapore's state-propagated narratives around its hoped-for future sustainability regimes. Instead, through this book and in the daily lives of those living in Singapore, it is ever more important for us to consider other stories to live by: stories that embody *beneficial* discourses in Singapore's cityscape and that make *salient* non-humans, degrowth, climate justice, community resilience, and strong sustainability. As we shared at the start of this chapter, the story of climate change and the internalized discourses that result from certain framings of it are key to a society's ability to respond, react, and reimagine society in relation to the natural and built environments. Despite the urgent harms of climate change and the framing of Singapore as an ecomodern, green, sustainable city, in writing this book, we have found it challenging to find truly *beneficial* discourses in our lived experiences of Singapore. This gap suggests that deeply sustainable narratives do not match with the current urban abstractions of our daily lives, where the human dependency on a planet not surpassing boundaries is *erased* by advertising, distant supply chains, and ecomodern promises, and where past and present colonial and gender violence are equally *erased*.

Such *erasure* is likely the case for many metropolises and their urban, non-nature areas that have been impacted by consumerist, globalized, and technocentric lifestyles the world over. However, there are many global and Singaporean examples of resilience, resistance, and persistence of deep environmental values, ones that do constitute *beneficial* discourses. Some of the Singaporean examples will be analysed in Chapter 5, where such places that embody these discourses point towards other possible pathways for Singapore – pathways aligned with the ecosophy of this book as articulated in Chapter 2.

In this chapter, we have explored a variety of narratives related to anthropogenic climate change. We have encountered science-based understandings; some of the political, economic, and social causes of climate change; narratives of lived realities (heat and natural disaster), especially of those living in Singapore; stories of climate politics and global (in)action; and encountered theoretical narratives (ecomodernism and metabolic rifts). Overall, these varied narratives help to contextualize various responses to climate change, as well as to other interconnected environmental issues. Each of these narratives contains different discourses and ideologies, feeding into various *stories-we-live-by*. It is hoped that these narratives provide an introduction to important concepts that ground the ecolinguistical analyses of Singapore that make up the core of this book.

In the chapters to follow, we offer specific ecolinguistical analysis and perspectives on Singapore, given the context of the climate crisis explored throughout this chapter. In Chapter 2, we also offer the philosophical and theoretical framework for such ecolinguistical analysis: our ecosophy. In this book's exploration of Singapore's *stories-we-live-by*, we attempt to map our lives into the larger global narrative of the climate crisis. In doing so, we hope that strongly sustainable *beneficial* discourses may be cultivated and grown from here onwards, whether in Singapore or elsewhere, that benefit all species in trying to navigate the catastrophic anthropogenic global warming yet to come.

Notes

1 We recognize that as of 2025, as this book is going to press, the Earth surpassed the 1.5°C for an extended period of time. This signals that this threshold was crossed, or at the very least will soon be crossed.

2 Historical emissions refers to greenhouse gas emissions made decades or centuries prior, yet that are responsible for a portion of the current excess in carbon emissions that the Earth's systems can handle without triggering extreme climatic change.

3 Singapore is caught in both trajectories: it is shaped by dispossession and colonialism, then by rapid industrialization and 'greening'. Yet Singapore is economically powerful in the region and has huge per capita emissions.

4 See especially Chapter 10.
5 These include securing investments in biofuels and manufacturing. For more information, readers should refer to https://www.nccs.gov.sg/singapores-climate-action/clean-technology/.
6 For concerns specific to Southeast Asia, readers should refer to ASEAN's State of Climate Change Report that highlights the region's specific vulnerabilities and strategies (ASEAN Secretariat 2021).
7 See the below section on Singapore's ecomodernization for an exploration of metabolic rifts.
8 For more on unsustainable social and environmental externalities of fossil-fuel and corporation-based industrial urban lifeways, see Carson (1991) and Klein (2014).
9 Weak sustainability advocates for an equal balance of social, environmental, and economic systems, whereas strong sustainability privileges environmental systems first, then social systems as embedded within environmental systems, then economic systems (of which there can be many different possible types) as embedded within the other two systems. The nomenclature is not meant to disparage weak sustainability as being 'weaker' or not capable of reaching sustainability – rather, the term 'weak' instead is in relation to its perceived lack of taking environmental baselines and laws of thermodynamics as the starting point and requirement of any concept of sustainability. Weak sustainability is instead open to economic capital replacing environmental capital, as they are considered equal in worth. Alternatively, weak sustainability may also be considered more anthropocentric, while strong sustainability is more ecocentric.
10 Singapore provides an interesting case study because, as much as the government wants continued growth in the economic sphere, it also recognizes the need to protect ecosystems and to manage its economic growth in a way that is sustainable economically, socially, and environmentally.
11 Broadly, this refers to the observation that there is often an initial rise of environmental pollution and harm with the onset of any industrial technology, but it is usually followed by a plateau, then a decrease in pollution and harm as efficiencies and regulations kick in to manage the technology.
12 The direct quote and summary of their view on decoupling are from https://www.ecomodernism.org/. Accessed 11 June 2024.

References

Aik, Joel, et al. 2023. 'Climate Change and Population Health in Singapore: A Systematic Review'. *The Lancet Regional Health Western Pacific* 40: 100947.

Asafu-Adjaye, John, et al. 2015. *An Ecomodernist Manifesto*. Self-published at https://www.ecomodernism.org/.

ASEAN Secretariat. 2021. 'Asean State of Climate Change Report (ASCCR)'. Jakarta. https://asean.org/wp-content/uploads/2021/10/ASCCR-e-publication-Correction_8-June.pdf.

Berastain, Pierre. 2024. 'Gender, Heat, and Cities: Building Resilience for a Changing Climate'. *Centre for Public Impact* (blog). Accessed 25 November 2024. https://centreforpublicimpact.org/resource-hub/gender-heat-and-cities-building-resilience-for-a-changing-climate/.

Bhullar, Lovleen. 2013. 'Climate Change Adaptation and Water Policy: Lessons from Singapore'. *Sustainable Development* 21 (3): 152–9.

Bonnedahl, Karl Johan, and Pasi Heikkurinen. 2018. 'The Case for Strong Sustainability'. In *Strongly Sustainable Societies: Organising Human Activities on a Hot and Full Earth*. London: Routledge.

Brozovic, Danilo. 2023. 'Societal Collapse: A Literature Review'. *Futures* 145: 103075.

Carr, John and Tema Milstein. 2021. '"See Nothing but Beauty": The Shared Work of Making Anthropogenic Destruction Invisible to the Human Eye'. *Geoforum* 122: 183–92.

Carson, Rachel. 1991. *Silent Spring*. London: Penguin Books (reprint).

Chen, Heather. 2024. 'Searing Heat Is Back across Southeast Asia and It's Not Going Away Anytime Soon'. *CNN*. Accessed 11 April 2024. https://www.cnn.com/2024/04/10/asia/southeast-asia-extreme-heat-climate-intl-hnk/index.html.

Daggett, Cara. 2018. 'Petro-masculinity: Fossil Fuels and Authoritarian Desire'. *Millennium: Journal of International Studies* 47 (1): 25–44.

Diffey, James, et al. 2022. '"Not About Us Without Us" – The Feelings and Hopes of Climate-concerned Young People Around the World'. *International Review of Psychiatry* 34 (5): 499–509. https://doi.org/10.1080/09540261.2022.2126297.

ECMWF. 2024. 'Copernicus: 2023 Is the Hottest Year on Record, with Global Temperatures Close to the 1.5°C Limit | Copernicus'. Accessed 9 January 2024. https://climate.copernicus.eu/copernicus-2023-hottest-year-record.

Ekers, Michael, and Scott Prudham. 2017. 'The Metabolism of Socioecological Fixes: Capital Switching, Spatial Fixes, and the Production of Nature'. *Annals of the American Association of Geographers* 107 (6): 1370–88.

Elkington, John. 2004. 'Enter the Triple Bottom Line'. In *The Triple Bottom Line: Does It All Add Up*, 16 p. London: Routledge.

Elmhirst, Rebecca, Carl Middleton and Bernadette Resurrecion, eds. *Migration and Floods in Southeast Asia*. London: Routledge, 2017.

Falkner, Robert. 2016. 'The Paris Agreement and the New Logic of International Climate Politics'. *International Affairs* 92 (5): 1107–25. https://doi.org/10.1111/1468-2346.12708.

Fletcher, Charles, et al. 2024. 'Earth at Risk: An Urgent Call to End the Age of Destruction and Forge a Just and Sustainable Future'. *PNAS Nexus* 3: 1–20.

Garekar, Bhagyashree. 2023. 'We Are Techno-Optimists by Nature: DPM Wong'. *The Straits Times*. Accessed 14 October 2023. https://www.straitstimes.com/world/united-states/we-are-techno-optimists-by-nature-dpm-wong.

He, Wenhui, Liqing Zhang and Chao Yuan. 2022. 'Future Air Temperature Projection in High-Density Tropical Cities Based on Global Climate Change and Urbanization – A Study in Singapore'. *Urban Climate* 42: 101115.

Heikkurinen, Pasi, Todd LeVasseur and Toni Ruuska. 2022. 'Imagining a Prosperous Periphery for the Rural in 2050 and Beyond'. In *Handbook of Sustainability Science in the Future*, 1–18. New York: Springer. https://doi.org/10.1007/978-3-030-68074-9_12-1.

Ingold, Tim. 2000. *The Perception of the Environment: Essays in Livelihood, Dwelling and Skill*. New York: Routledge.

Intergovernmental Panel On Climate Change (IPCC). 2023. *Climate Change 2022 – Impacts, Adaptation and Vulnerability: Working Group II Contribution to the Sixth Assessment Report of the Intergovernmental Panel on Climate Change*. 1st ed. Cambridge University Press. https://doi.org/10.1017/9781009325844.

IPCC and WMO, eds. 1992. *Climate Change: The 1990 and 1992 IPCC Assessments, IPCC First Assessment Report Overview and Policymaker Summaries and 1992 IPCC Supplement*. Geneve: IPCC.

Ipsun Solar. 2022. 'COP27, the Resilience Race, and Noteworthy Voices of Climate Leaders'. *Ipsun Solar* (blog). Accessed 15 November 2022. https://ipsunsolar.com/blog/cop27-the-resilience-race-and-noteworthy-voices-of-climate-leaders/.

Islam, M. Rezaul, and Niaz Khan. 2019. 'Threats, Vulnerability, Resilience and Displacement Among the Climate Change and Natural Disaster-Affected People in South-East Asia: An Overview'. In *Climate Change Mitigation and Sustainable Development*. London: Routledge.

Kemp, Luke, et al. 2022. 'Climate Endgame: Exploring Catastrophic Climate Change Scenarios'. *Proceedings of the National Academy of Sciences* 119 (34): 1–9.

Klein, Naomi. 2014. *This Changes Everything: Capitalism vs. the Climate*. Riverside: Simon and Schuster.

LeVasseur, Todd. 2021. *Climate Change, Religion, and Our Bodily Future*. London: Rowman & Littlefield Publishing Group, Inc.

LeVasseur, Todd, and Ewan Kingston. 2024. 'Are We Doomed? A Climate Conversation'. *Cosmos and History: The Journal of Natural and Social Philosophy* 20 (2): 1–32.

LeVasseur, Todd, and Bernard Zaleha. 2019. 'When Christ Is Maladaptive'. In *Christian Theology and Climate Change*, 120–31. New York: T & T Clark/Bloomsbury.

Magramo, Heather, and Kathleen Chen. 2024. '"Blast-Furnace Heat Every Day": Record Temperatures Cancel Classes, Widening Learning Gaps across Southeast Asia'. *CNN*. Accessed 9 May 2024. https://www.cnn.com/2024/05/09/asia/southeast-asia-heatwaves-education-school-closures-intl-hnk/index.html.

Mann, Michael E., Raymond S. Bradley and Malcolm K. Hughes. 1999. 'Northern Hemisphere Temperatures during the Past Millennium: Inferences, Uncertainties, and Limitations'. *Geophysical Research Letters* 26 (6): 759–62. https://doi.org/10.1029/1999GL900070.

Marthin, Anggita, and Louis Budiman. 2020. 'The Discourse of Climate Migration: Unravelling [sic] the Politics of ASEAN's Environmental Policies'. *Pacific Journalism Review* 26 (2): 35–51.

Merz, Joseph, et al. 2023. 'World Scientists' Warning: The Behavioural Crisis Driving Ecological Overshoot'. *Ecology and Environmental Sciences* 106 (3): 1–22.

Meteorological Service Singapore. 2024. *Singapore Climate 2023: The Year in Numbers*. Singapore: Meteorological Service Singapore.

Ministry of Education. 2019. 'Environmental Education'. Parliament Replies. Last Modified: Jan 15, 2019. http://www.moe.gov.sg/news/parliamentary-replies/20190115-environmental-education.

National Climate Change Secretariat (NCCS). 2024. 'Clean Technology'. https://www.nccs.gov.sg/singapores-climate-action/clean-technology/.

OECD. 2015. *The Metropolitan Century: Understanding Urbanisation and Its Consequences*. OECD. https://doi.org/10.1787/9789264228733-en.

Ray, Sarah Jacquette. 2020. 'Gen Z: The Climate Generation'. *University of California Press* (blog). https://www.ucpress.edu/blog-posts/49184-gen-z-the-climate-generation.

Richardson, Katherine, et al. 2023. 'Earth Beyond Six of Nine Planetary Boundaries'. *Science Advances* 9 (37): eadh2458. https://doi.org/10.1126/sciadv.adh2458.

Rockström, Johan, et al. 2009. 'A Safe Operating Space for Humanity'. *Nature* 461 (7263): 472–5. https://doi.org/10.1038/461472a.

Roy, Natalie M. 2021. 'Climate Change's Free Rider Problem: Why We Must Relinquish Freedom to Become Free'. *William & Mary Environmental Law and Policy Review* 45 (3): 821–56.

Seric, Adnan, and Siong Tong Yee. 2019. 'What Are Global Value Chains and Why Do They Matter?' *Industrial Analytics Platform*. August 3, 2019. https://iap.unido.org/articles/what-are-global-value-chains-and-why-do-they-matter.

Singapore Government. 2016. 'Singapore's Climate Action Plan'. Accessed 12 December 2024 https://www.mse.gov.sg/resources/climate-action-plan.pdf.

Singh, Sabina. 2021. 'Green Colonialism'. *Fourth World Journal* 21 (2): 117–22.

Sultana, Farhana. 2023. 'Whose Growth in Whose Planetary Boundaries? Decolonising Planetary Justice in the Anthropocene'. *Geo: Geography and Environment* 10 (2): e00128.

Thompson, Paul. 2016. 'The Many Meanings of Sustainability: A Competing Paradigms Approach'. In *Pragmatic Sustainability: Dispositions for Critical Adaptation*, 16–28. New York: Routledge.

UNFCCC. 2015. 'The Paris Agreement'. Accessed 12 December 2024. https://unfccc.int/process-and-meetings/the-paris-agreement.

United Nations. 2018. '68% of the World Population Projected to Live in Urban Areas by 2050, Says UN'. *United Nations*. https://www.un.org/en/desa/68-world-population-projected-live-urban-areas-2050-says-un.

Wilson, Sheena, Imre Szeman and Adam Carlson, eds. 2017. *Petrocultures: Oil, Politics, Culture*. Quebec: McGill-Queen University Press.

WMO. 2024. 'WMO Confirms That 2023 Smashes Global Temperature Record'. World Meteorological Organization. Accessed 11 January 2024. https://wmo.int/media/news/wmo-confirms-2023-smashes-global-temperature-record.

Zografos, Christos. 2022. 'The Contradictions of Green New Deals: Green Sacrifice and Colonialism'. *Soundings* 80 (80): 37–50.

1

The Singapore *Stories*

Anjali Venketram, Beverley Choo Jia Ying,
Hannah Ng Shueh Yi, Kristina Gweneth Ponce Simundo, Lecarl Lim
and Todd LeVasseur

Introduction

As explained in the preface and introduction, understanding the narratives, perspectives and discourses that shape modern urban realities offers a possible analysis of their unsustainability and a direct vein for any person to intervene in these concerning trajectories. This is because stories are malleable and changeable by the people living in, constructing, and telling them. In the ecosystem of narratives that make up the fabric of our modern experience of cities, we turn our gaze to Singapore, a tiny island city-state in Southeast Asia with a land area of 734.3 kilometres[1] residing at an average of 15 metres above sea level.

Like other city and urban-dwellers around the world, the designed and built environments of Singapore create the discourses and *stories-we-live-by* in the mind of every Singaporean. As explored in the preface, whether conscious or not, each person has internal beliefs about the more-than-human world. These beliefs in turn help to create and manage the materially designed and engineered environmental 'texts' within which people dwell; texts that are based on the discourses, and key, the ideologies those discourses contain, and that people thus believe in and act out. It is therefore critical to understand what type of stories and discourses have contributed to creating the cities we live in, laud and hope for, both in Singapore and any other urban area the world over.

While relatively small compared to other major cities in the world, the stories told within and through Singapore are ones that reflect significant themes of global modernity and its treatment of the ecosystem.

The 'Singapore Story': A strong city-state, capable of navigating the colonial to post-colonial period of the region during the era of post-World War II nation-building; and now, equally capable of navigating a global economy as a formidable force, as an 'Asian Tiger'. Singapore is consistently lauded as a 'poster child' for a city heading towards climate survival and ecomodern thriving within emerging 21st-century socioenvironmental metrics. The main narrative told about Singapore is its global recognition as an ecomodernist paradise – in a cursory internet search of the city-state, images of clean streets, efficient waste management and neat blocks of public housing appear. The quintessential 'Singapore Story' is a tale of urbanization and modernization through decades of rapid development transforming a 'fishing village' to the city we (outside of Todd) all currently live in.

The Singapore Story is the dominant narrative of Singapore's history of post-independence development, one that is state-centralized (Afandi and Lim 2022) and standardized across all of Singapore's public schools. This version of the Singapore Story is largely written by the ruling party and is the dominant story about Singapore to which people around the world are exposed. This story has also shaped the country's environmental managerialism, its self-understanding and its economic and thus industrial development.

Given the focus of this book, and Singapore's reputation as a green city of the future, it is the sustainability and ecomodern aspects of its popular story on which we focus. This exploration builds on the book's introduction and its analysis of the ethical, ecological, structural, economic and political problems facing cities in the twenty-first century and beyond, especially from sustainability metrics. The introduction allowed for a more comprehensive and critical analysis of Singapore to occur here but that should be read within that part of the book's established broader context.

In this chapter, we aim to recount this dominant Singapore Story, then move on to focus on telling different sides of this story. All recounts of this Singapore Story are from the perspective of critical residents looking outwards at decades of shifting global developmental trajectories – and Singapore's national narratives (today and historically) in the face of them. Considering the developmental trajectories and struggles of nearby Southeast Asian countries, Singapore offers an interesting case study within the region to examine the flows of power, capital and logics that underlie this country's utopian image of sustainability. The following narratives will provide us with the necessary foundations to subsequently understand the critical ecolinguistic analysis we will engage in for the main bulk of this book.

Telling Singapore's Story (or rather, Stories…)

Thus, in the following sections, we explore how Singapore's national values with regards to its treatment of nature have changed over time, responding to changing political environments and national priorities in line with the government's vision and pursuit of modernity. These values have manifested themselves in three key narratives:

1. From Fishing Village to Metropolitan City;
2. Paternalistic Development; and
3. The Green City.

Narrative One: From Fishing Village to Metropolitan City

Among the first things that any resident of Singapore learns about the Singapore story is the arc of its development 'from fishing village to metropolitan city'. The contrast of the image of its almost unimaginable rural beginnings and its present-day largely urban, built-up area *is* the Singapore story – a narrative of growth, development and change. In this section, we chart that journey of rapid change, while also filling in gaps that are oftentimes missed out on in Singapore's story of the last few centuries. We note that understanding this journey is integral as context for the hopes and dreams we have for Singapore moving forward, further outlined in our Chapter 2: Ecosophy.

The image of the 'fishing village', often referred to as Singapore's 'beginning point', is a pre-colonial fantasy of an underdeveloped, barren, rural land. The story of pre-colonial Singapore remains largely underdiscussed in mainstream tellings of the Singapore story. However, as many scholars of Singapore have written to fill in the gaps, we now know this wasn't quite the full picture of the island at that time. Singapore had a thriving indigenous population and a multicultural society a long time before the British arrived at the port. As previous Assistant Professor of Environmental Studies at YNC Matthew-Schneider Mayerson explains,

> The delimited sovereign territory composed of solid landmass and azure ocean that we now refer to as 'Singapore' has been a meeting place for a wide variety of human and nonhuman beings since the Stone Age. They include countless generations of Orang Laut – the indigenous people of Singapore – as well as Srivijayan royalty, Siamese and Vietnamese traders, Chinese merchants, Indian coolies, Bugis, Chinese and Malay traders, British colonisers and modern-day,

> multicultural Singaporeans, along with hundreds of thousands of migrant labourers from Malaysia, the Philippines, Myanmar, Bangladesh, Indonesia, China and elsewhere. Due to its geographic location, mobility has long been central to the residents of the island. (2020, 13)[2]

What changed between then and now that has resulted in such fundamental shifts in mindset among Singapore's residents? In the decades that we will cover, the narrative of the 'fishing village to metropolitan city' tells a story of changing Singapore's residents' relationship to the land and, key, their relationship to capital. The start of those mindset changes began with the advent of colonialism in Southeast Asia. This colonialism perpetrated a 'tragedy of enclosure' (Ghosh 2021) throughout the entirety of Southeast Asia. Through colonization, Europeans violently rearranged lands historically partly held in commons into various colonial enterprises based upon dispossession, hierarchy, class privilege, caste, European technologies and weapons, European modernity, export Green Revolution agriculture, the demands of European markets and consumers and European imperialism.

The Portuguese came to Singapore in the 1500s, attracted to its geographic location and conditions – they recognized the relatively low tidal shifts, comparatively gentle currents, tropical equatorial weather, dependable monsoon winds and a historically rich biodiverse island geography. However, by the early 1600s, the island was abandoned by colonial machinations, reverting back to Indigenous population and control, and therefore to a continuance of the trade and migration networks that existed pre-colonization. Malay officials settled on the island in 1818, and Chinese labourers and planters began to arrive in earnest as well. The 1800s was when the British came to Singapore, beginning the bulk of the narrative of rapid change in Singapore. Sir Thomas Stamford Raffles disembarked onto the island in 1819, representing the British East India Company, who were granted the rights to operate a trading post by the Sultan Hussein Shah of Johor on 6 February 1819. Among the major transformations the British enacted included changes in land use for capital gain. The British, via deforestation and mass land-use conversion, reformed Singapore into a plantation economy of pepper, rubber, nutmeg, gambier and other export crops.[3]

Over time, Singapore began to be strategically established as a significant port in the region (as it is still known today[4]). Its port-of-call is a key aspect of its narrative of success and development. Singapore became increasingly important to regional trading states due to its strategic location along key trading lines, bringing more workers to the island from surrounding countries in the region.

A lot of Singapore's infrastructure and labour force at the time consisted of migrant labour. This influx of migration as labour continues today, albeit in a different form – hard labour and domestic work largely depend on migrant workers from nearby Global South countries. Overall, the geopolitical as well as economic importance of the port made the British reluctant to cede their control over Singapore, though they eventually did in the 1950s. Singapore merged with the Federation of Malaya to form Malaysia in 1963 but was expelled because of social unrest, racial tensions and political differences. The Republic of Singapore was then formed on 9 August 1965 (Turnbull 2009).

In the wider arc of the 'fishing village to metropolitan city' narrative, a key story is one of struggling through uncertainty, to attain the stability Singapore has today. Through its newly formed independence, Singapore's residents struggled through tensions from an era of nation-building, navigating its colonial and imperial histories. This uncertain era of nation-building in Singapore and wider Southeast Asia saw colonial and imperial histories superseded by globalization and tourism. Singapore navigated this by maintaining a strong narrative for its early years as a young nation, steering it towards multi-ethnic harmony and economic growth.

Early-independence Singapore in the 1960s was, like many other post-colonial settlements, one that suffered from a lack of modern infrastructure: lacking in running water, sewage or waste treatment, government-supported educational systems, modern healthcare, adequate housing and a land environmentally decimated by its plantation past. The homes in traditional kampongs[5] were given a terrible reputation and were classified as slums and squatter settlements with poor sanitation and unhygienic conditions (Housing Development Board 2024). Early government housing policies, therefore, aimed at remedying these perceived shortfalls of what a modern nation should have. Peter Ho posited that the housing slums in Singapore were located on 'valuable land that could be used for commercial development' (Guo 2016, xi), and thus, mass resettlement of residents into Housing Development Board homes ensued.

These efforts of provision of basic needs and services were largely successful and helped to solidify decades of support from Singaporeans for the dominant party at the time – the People's Action Party (PAP). Apart from housing, the early government of Singapore set up a Public Utilities Board (PUB) to develop, operate and maintain water municipal needs, and it created the National Environment Agency (NEA) which oversees waste management and environmental issues and many other such governmental bodies which continue to play a pivotal role in maintaining the high quality of life Singapore's residents enjoy.

Thus, to this day, this same party remains the incumbent and hegemonic party in Singapore, maintaining over 80% of parliamentary seats since its first election in 1959. Hence, the PAP is a key character in this part of the narrative – the PAP has been integral to Singapore's struggle through uncertainty, and a foundational institution to its stability and success today. As a result, it maintains its hegemonic power in Singapore's sociopolitical imagination, its governing bodies and institutions and its national decision-making today.

Equally key to the Singapore story as the narrative of struggle is the narrative of scarcity. Singapore's governance in the early years of its nation-building (and even largely philosophically underpinning its decisions today) aimed to grow and develop the country. In a literal sense, a part of this meant growing its land mass. Singapore is one of the densest places on the planet in terms of humans per square-kilometre,[6] and because it has a limited land mass, land scarcity became a central concern and challenge that the PAP would quickly begin to face in terms of long-term planning for sustainability (Fujii and Rohan 2019).

Land reclamation was a big part of Singapore's growing industry and manufacturing in the 1970s and 1980s. Such reclamations transformed what was a collection of islands at the bottom of Southeast Asia, into an 'island nation' – where this diversity and history created Singapore's 'history of streets, highways, railways, farms, factories, docks, islands, shop houses, kampongs, shanty towns, bungalows, slums, high-rises, mosques, temples, churches, schools and ethnic enclaves' (Barr 2014, 173). This process of reclamation included removal and relocation of the indigenous Orang Laut peoples from many of Singapore's offshore islands,[7] leading to a rapid loss of traditional, place-based ecological knowledge, as well as the loss of many coastal, intertidal and marine habitats. Such reclamation is a major component of Singapore's plans for its continued geographic and economic growth, as well as part of its plans for adapting to sea-level rise that is driven by human-induced global heating.

Apart from land scarcity, a common narrative line in Singaporean political discourse is of Singapore, 'a small island city-state with limited natural resources' (Singapore's Voluntary National Review n.d.). Its limited land area, coupled with extensive land-use conversion to high-rise housing or industrial buildings, has left Singapore with little to no natural resources for consumption or export, thus making it heavily reliant on the import and export of goods. This narrative is instead converted into a discourse and practice of intentional people management – Lee Kuan Yew, often credited as the founding father of Singapore, and his government, invested heavily in the education and development of its people – 'Singapore's only available natural resource' (Yiannouka 2015). The strategies involved in this development are further explored in the next section, Narrative Two.

Through the narrative covered in this Singapore Story, a heavy emphasis is made on the leaps and bounds made by the city-state, but often overlooked are the costs associated with these gains. For example, the labour required to build and maintain the metropolis of Singapore depends on foreign labour who are excluded from most benefits: healthcare, housing with access to climate control, privacy and even safe modes of transport to and from work sites. For example, construction workers are typically ferried in open-air lorries which are otherwise acknowledged as an unsafe practice. Yet such foreign labour forms approximately 40% of Singapore's workforce. In 2024, nearly half of all workplace deaths in Singapore were in the construction sector, a sector primarily made up of migrant workers (Lim 2025a).

A critical reflection, especially from a strong sustainability perspective, on the modes of development in this dominant 'Singapore Story' narrative is thus warranted. Besides the (migrant) human cost of development, the rapid development of Singapore from the 1800s onwards has decimated most of its natural ecosystems while also alienating much of its population from nature. These dynamics are further explored in the following narratives. The next one expands on Singapore's paternalistic modes of development, the sustainability of said development and the resulting solastalgia because of the alienation from nature that is part of Singapore's paternalistic governance.

Narrative Two: Paternalistic Development

In the prior section, we briefly explored the evolution in which Singapore has conceived of its history. Crucially, however, we must note that this imagination has been led by the state. The developmental trajectory as outlined in Narrative One very much depended on the control of the government, creating the paternalistic governmental narrative still dominant in many facets of daily Singaporean life today. Indeed, Singapore has been identified by many as a 'developmental state' (Liow 2012; Low 2001; Siddiqui 2016) that engages in 'authoritarian environmentalism', which will be explored further below (Han 2017). The government has had a heavy hand in shaping the country's economy, culture and planned environment which by standard developmental metrics has been successful.[8] The structure of the civil service is led and supported by a political system designed to retain its power and influence over the direction of Singapore's progress. According to sociologist Chua Beng Huat,

> After close to six decades of continuously governing Singapore, the PAP government is now in its transition to a fourth generation of political leadership, and it has developed a very stable political system. This system features a decidedly anti-liberal social management system that places social interests above individual rights and freedoms and an economy that is a mix of state-owned and private capital. (Chua 2017, 193)

Chua's analysis continues, noting that,

> Everyone acknowledges the undeniable fact that the material life of the entire nation has improved massively under more than half a century of PAP rule as a single-party dominant government. Everyone is also acutely aware of the equally undeniable fact that this improvement has been paid for with a significant quantum of social, cultural, and political repression, strewn across the political history of the island nation, particularly in the first 30 years under Prime Minister Lee Kuan Yew [and even the PAP] would insist that repression was necessary to produce an improved material life for all. (Chua 2017, vii)

The political strength of the PAP continues through the time of writing, as seen in the General Elections of 2025, where, despite a seemingly strong presence of a diversity of opposition candidates, the PAP won in a 'landslide victory' (Lim 2025b), securing a super majority in parliament and a 65.57% vote share. Given such past (and present) numbers, Chua opined that Singapore's 'political sphere has not made any concession to ideological pluralism' (Chua 2017, 194). In the face of the geopolitical uncertainty and climate-changed futures as its backdrop, Singapore's continued dominant narrative is one of long-lasting trust in a paternalistic government despite the repression that such governance comes with. The fixation on material growth through economic stability and of recognized repression are dominant narratives in Singaporean *stories-we-live-by*.

The modus operandi of Singapore's development geared towards economic growth as helmed by the PAP government relies on an extremely efficient system of policy making. Political office holders are supported by a civil service ordered by statutory boards, ministries and specially assembled task force teams that tackle planning. Environmental governance is no different, where environmental outcomes related to land-use changes and climate change adaptation policies are centrally controlled with limited avenues for engagement with civil society. Authoritarian environmentalism is thus a plausible descriptive model for Singapore's environmental governance.

Using political scientist Bruce Gilley's definition, authoritarian environmentalism can be provisionally defined as a public policy model that concentrates authority in a few executive agencies manned by capable and uncorrupt elites seeking to improve environmental outcomes. Public participation is limited to a narrow cadre of scientific and technocratic elites, while others are expected to participate only in state-led mobilization for the purposes of implementation. The policy outputs that result include a rapid and comprehensive response to the issue and usually some limits on individual freedoms (Gilley 2012, 288). Gilley is quick to acknowledge, however, that in practice, 'all policy models are mixtures of democratic and authoritarian features' (2012, 289). He suggests two metrics of participation in democratic environmentalism (which authoritarian environmentalism lacks) measured by the stage and level of participation – referring to up or downstream stages in decision-making and low to high levels of participation (2012, 289).

In Singapore's case, due to its developmental trajectory, spatially fixed industrial investments and historical environmental authoritarian decision-making, civic participation has been selective, though current-day policymaking processes typically involve engagement of civic actors by government agencies. This, alongside the already established image and history of Singapore as the clean, 'green', garden city continues to put forth a highly impersonal, abstracted, manicured and controlled version of participation with issues related to its management of the environment: '[T]he government attempts to truncate environmental concerns by interweaving the environment into discourses of nation-building, leisure, heritage, efficiency and economic survival' (Hobson 2005, 160). Hobson's 2005 analysis of environmental non-governmental organizations (NGOs) concluded that environmental NGOs in Singapore operate in the paradoxical cycle of governmental influence, where sociopolitical relations between them and policy makers/decision makers in the government are more flexible, informal and semi-formal and allow space for negotiations but operate without assurance that real impact can be made, therefore resulting in disenchantment (Hobson 2005, 171).

Despite being written close to two decades prior, Hobson's analysis continues to ring true, especially as experienced by those of us writing this book who grew up in Singapore. Singapore's trade-off is between lower levels of true inclusion and participation, of community (NGOs and other humans) and non-humans in decision-making (ala Gilley) and of efficiency and control over developmental direction and speed.[9] Since Hobson's 2005 analysis, Singapore's environmental authoritarian approach has evolved – public engagements on developmental and policymaking decisions are required, though the form and impact vary.[10]

In 2013, Tan Chuan-Jin, former Minister for Manpower and previous Senior Minister of State for National Development, even went so far as to saying,

> The government ought to see engagement as an important process, but it is not a blank cheque to engage to death. As a leader, your job is to create consensus if you can. If not, at some point you have to make choices and some of these aren't popular, aren't easy… But fearing you're not getting it right shouldn't freeze you from action because I think that is a dereliction of duty. (Kong 2013, 13)

The implication here is that Singaporeans are only consulted as long as their opinions do not drastically slow down the processing time for the government. Repression is thus a key facet of any analysis of Singapore as a sustainable city.

Apart from repression of free political speech, systemic methods of environmental and social repression exist as well. For example, this repression often has been of Indigenous and Malay peoples at the expense of ethnic Chinese peoples in Singapore (Chua 2009) and is especially seen in the relocation and thus dispossession of the Orang Laut Indigenous peoples from most all of Singapore's many smaller islands (for a brief example of this, see Chapter 5's case study on Chek Jawa). As the political geographers Creighton Connolly and Hamzah Muzaini explain, 'land reclamation produces a form of socio-ecological violence that [is] highly uneven, affecting particularly Singapore's Malay and indigenous *orang laut* communities that have been displaced for development' (2022, 2173).

There are additional layers of environmental and ecological repression here, apart from its indigenous peoples, in that the ecosystems of those same islands are literally paved over in the active pursuit of material gains and abundance. Connolly and Muzaini, in an analysis similar to that taken in this book, explain that 'Singapore and its offshore islands are [thus] illustrative of the perpetual churning of socio-spatial formations under capitalism and the numerous metabolic flows involved' (2022, 2177).

As the one-time columnist of *The Straits Times* Cherian George explains, rather than the Singapore Story as Narrative One covered, instead what is needed is to 'think of Singapore instead as the air-conditioned Nation – a society with a unique blend of comfort and central control, where people have mastered their environment, but at the cost of individual autonomy, and at the risk of unsustainability' (2010, 15). Given Singapore's 2025 overshoot day was February 26,[11] our current metrics of consumption continue to be unsustainable.

The Singapore government's decision to run the island nation as a developmental state is a crucial one – it enabled the 'Fishing Village to Metropolitan City' narrative and gives the state disproportionate power in perpetuating this narrative and shaping other *stories-we-live-by*. When approaching our natural environment, this has translated into authoritarian environmentalism. In response to the rapidly worsening climate crisis, Singapore continues to brace itself for inevitable environmental impacts through top-down coastal and flood resilience plans that advocate technology-based ecomodernist solutions (Public Utilities Board 2023). The next section will explore and situate such trajectories for Singapore in a more sustained way, highlighting what Singapore does well from a sustainability perspective and where it falls short.

Narrative Three: The Green City

The 'first ever tree' planted in Singapore was the symbolic 'start' of the nation-state's movement towards greening the country. The tree was planted on June 16, 1963, by Singapore's first Prime Minister Lee Kuan Yew.[12] It represents a snapshot of Singapore at that time – a country newly independent from British colonial rule and beginning to forge the first roots of its national identity. The Singapore Story spirals forward from the planting of the 'first tree' in 1963, towards a larger narrative curating Singapore to be an exemplar of human–nature relationship. On 11 May 1967, then Prime Minister Lee Kuan Yew famously championed the vision of Singapore as a 'garden city'. Prime Minister Lee's garden city vision has resulted in millions of trees being planted throughout Singapore and has contributed to the narrative of Singapore being a 'garden city':[13] the blossoms of the Singapore Story seed planted in Lee's 1967 speech have since matured into the fruits of a global discourse that centres Singapore as a 'sustainable' city of the future, seen internally and externally as a literal and figurative 'green' city. As will be explored shortly, this is especially true for many who are not familiar with the actual 'green' reality of contemporary Singapore that this book will explore, challenge and celebrate.

The Singapore of the twenty-first century builds upon Prime Minister Lee's 1960s and 1970s evocative vision of the multi-ethnic and multiracial (Kuo and Chan 2016) garden city that he planted last century. This emerging vision of Singapore was carried on by Lee Kuan Yew's son, then Prime Minister Lee Hsien Loong,[14] who on 28 June 2012 during the opening ceremony for the Gardens by the Bay where, building upon the discursive rhetoric of his father, the younger Lee espoused that 'The next phase from being a "Garden City" is to make ourselves into a "City in a Garden" [that] means connecting our communities and our places and spaces through parks, gardens, streetscapes and skyrise greenery … And we are going to bring the green spaces and the biodiversity closer to our homes and workplaces' (Lee 2012).[15] The phrase, City in a Garden, is a short, eloquent and evocative revisioning of what an economically powerful, multi-ethnic city, ASEAN[16] city, aspires to be in the twenty-first century. While building on Lee's father's legacy, the new phrase linguistically decentres the human, yet also implies a manicured, landscaped garden (more will be spent on this *framing* in Chapter 4). This Singapore Garden evokes an image of abundance, health, pleasure and aesthetic beauty, where Singapore as a city absorbs these same attributes via association with and location within the garden. A key result of this *framing* is that Singapore is held up as an international leader in creating 'biophilic urbanism' (Tabb 2021). The current iteration of this narrative from the Garden City to the City in a Garden is the 'City in Nature',[17] a vision shared by the National Parks Board (NParks) of Singapore published in 2020 'as part of the next bound in urban planning, which aims to ensure a green, liveable and sustainable home for Singaporeans for generations to come' (Er 2020, 4). This 'City in Nature' is further discussed in Chapter 4 as a standalone case study.

A constant theme running through these autobiographical framings of Singapore by its leaders is an understanding of a certain relationship with nature and therefore the place of the city within nature. This *frame* is that Singaporeans can benefit in various ways from nature, while designing a high-tech ecomodern city of the future. An example of a *framing* of Singapore as a clean, attractive garden is seen in the Foreword written by then President of the Republic of Singapore, Dr. Tony Tan Keng Yam, in the internally published 2013 National Parks (NParks) Publication, *Living in a Garden: The Greening of Singapore* (2013). The Foreword reads

> Singapore is recognised today for being a clean and green city. Our lush landscape is an important facet of our identity. The garden environment and green spaces contribute to an enjoyable and liveable environment in which Singaporeans live, work and play. Visitors to Singapore are struck by the sight of tree-lined highways, manicured lawns and smartly-pruned hedges that greets them. Our reputation as a City in a Garden enhances Singapore's attractiveness as a destination for tourists, foreign businesses and global talents. The many vibrant parks and gardens, the biodiversity of our nature reserves, and the extensive network of park connectors and green corridors that we now enjoy

> are our natural heritage. These are the results of deliberate and sustained efforts that began 50 years ago.

This dominant *framing* is part of the internal and external narrative of Singapore as a green, arboreal city of the future, well-rooted in visions from the past. After the end of the British mandate in Singapore, the nation was grappling with the need to create a strong identity to establish itself on an international platform. The 'clean and green' narrative and framing helped (and still helps) Singapore market itself to tourists and investors, while championing itself as a sustainable, ecomodern green city of the future. It has been instrumental in justifying Singapore's claim to uniqueness in the region in sustainability, contributing to the claimed identity of a Singapore whose boundaries blur between nature and urban.

Another *framing* related to the sustainability of Singapore can be seen with the architect Puay Yok Tan and members of the National Parks Board of Singapore James Wang and Angelia Sia's fifty-year analysis of Singapore's greening efforts. In this cumulative analysis, they point out that despite the city nation's increasing population density and ageing tree populations that lack full biodiversity of flora types, Singapore nonetheless 'embarked in an active urban greening programme early in its urban developmental history with an emphasis on creating a clean and green city, before issues of quality of the physical environment have come to the forefront of current discussions on cities' livability and environmental sustainability' (Tan et al. 2013, 30).

Such celebration and recognition are also seen in the summary of metaphors about Singapore as garden: from 'garden city', to 'city in the garden', to 'rainforest in a city' (Chua 2015), to the even more recent government plans of NParks that tout Singapore as a City in Nature. Here, Peter Rowe, professor of Architecture and Urban Design, and Limin Hee, Director of Research for the Centre for Liveable Cities in Singapore, point out that,

> The trend from Garden City to a City in Nature mirrors the circumstances of Singapore during [colonial to independence] time. On the heels of vast destruction of natural and garden-like circumstances, as well as rounds of cleaning up, bringing the garden into the city would seem to follow both logically and practically... Furthermore, as the Housing and Development Board's program gained momentum and success, the sheer need to offset the built and constructed environment with something more natural, and no doubt softer, became more and more obvious. Then because of the persistence of these efforts [tree planting beginning in the early 1970s], Singapore became tangibly

> greener to the point where the scale tipped, as it were, to Singapore being able to be perceived as a garden more broadly into which city building was occurring. (2019, 53)

In research and academic literature produced in Singapore, the greening of the city has always been associated with positive effects and has historically and collectively embedded the idea of greening as a positive effort by Singapore – which claims to have nothing but benefits for the people of Singapore – whether benefits be economic, or in terms of physical and mental liveability.

However, not all of Singapore's green initiatives are driven and directed by top-down government action. Indeed, there is a strong civic dimension to Singapore's green identity, too (a theme that will be expounded upon in a later section and in Chapter 5). For example, ecotourism scholar Aaron Tham did a case study of three ecotourism spots in Singapore – Sisters' Island Marine Preserve, Kampong Kampus Ground-Up Initiative, and KEP, a kayaking ecotour outfit affiliated with the Waterways Watch Society – and found that 'the catalyst for proposed ecotourism ventures [in Singapore were] derived with almost minimal government involvement. Instead, the desire to undertake ecotourism was driven by the presence of individuals exhibiting a vision of sustainability and championing a cause' (Tham 2019, 12).

Despite such citizen-led initiatives, the dominant narrative being analysed in this section still centres the government of Singapore in creating a sustainable city in nature. This is seen for example with Singapore's Green Plan 2030, released by the Government of Singapore in February of 2021. The Green Plan 2030 is a 'whole-of-nation movement to advance Singapore's national agenda on sustainable development' (Ministry of Sustainability and Environment et al. 2021) aiming to achieve net-zero emissions in Singapore by 2050. Comprising five pillars, the plan outlines steps that aim to increase natural spaces,[18] attempt to 'green' Singapore's energy mix, transform Singapore into a Zero Waste Nation, decarbonize Singapore's industrial energy and chemical industries (read: petrochemicals and others) and strengthen resilience to sea-level rise and food security. Despite lauding Singapore's sustainable trajectory through this plan, it ultimately fails to address the contradictions that urban centres are prey to – as mentioned in the introduction, their density and urban industrial metabolism require a much larger ecological/land footprint (Global Footprint Network 2025).

Every focal pillar in the Green Plan 2030 requires resources or technologies that aim to reduce impacts, but these do not reduce production, consumption, or demand, making this at best a pathway towards weak sustainability. For example, the Green Economy pillar lays out 2030 targets as:

- Jurong Island is to be a sustainable energy and chemicals park;
- Singapore is to be a sustainable tourism destination;
- Singapore will become a leading centre for green finance and services to facilitate Asia's transition to a low-carbon and sustainable future;
- Singapore will be a carbon services hub in Asia;
- Singapore will be a leading regional centre for developing new sustainability solutions; and
- Singapore will invest in and mentor a strong pool of local enterprises in order to capture sustainability opportunities.

These aspects of Singapore's Green Economy pillar embrace technological and green growth solutions, capitalizing on financial gains from climate change trends towards a 'low-carbon and sustainable future' to transform the world's ninth largest chemical exporter as of 2022 into a 'sustainable, highly productive hub for chemicals manufacturing' (Ministry of Sustainability and Environment et al. 2021).[19] Even without further analysis here, it is apparent that Singapore is concerned with sustainability, at least in its weak manifestation. All of its sustainability efforts, however, are marred by the underlying logic of technocracy and ecomodernism which widens the metabolic rift between Singapore the city and Singapore the consumer; between Singapore the 'City in Nature' and the earth itself.

These pillars – and Singapore's approach to sustainability as a whole – can also be viewed through the lens of green capitalism. Proponents of green capitalism believe that 'innovation and technological progress can save the environment and still facilitate capital accumulation and ever-lasting growth in production' (Ponte 2023, 102). Many of Singapore's efforts to mitigate and adapt to climate change are fundamentally profit-driven (often euphemistically called 'market-based' and 'industry-driven'). From green finance and carbon markets to sustainable tourism, these efforts promise that we can solve our environmental dilemmas while still pursuing economic growth. Fundamentally, they maintain and fail to acknowledge (let alone question) the political economy that undergirds the socio-ecological crisis (Goldstein and Tyfield 2017).

Ultimately, despite the appearances of greenery, there is a shadow side of unconscious and often unacknowledged aspects (Plumwood 2008) to the creation and development of Singapore. The state's paternalistic and developmental state approach has manifested itself as authoritarian environmentalism in the face of climate change and a global push for sustainability. Then, the environmental strategy chosen under this governing style is fundamentally ecomodernist and aligned with green growth while largely avoiding addressing deeper systemic issues. In the unfolding Anthropocene, we need new understandings of human–human and human–non-human interactions (Schneider-Mayerson 2020). Yet, Singapore's current policies and ambitions fall short of this, to say the least.

In the next section, we will explore how the state's technocratic and ecomodern approach has impacted the experiences of Singapore's residents, their relationship with the local environment and the resulting stories-they-live-by.

The Need for Other Singapore Stories

The above narratives have helped to situate the state-centralized narrative about the Singapore Story, its developmental trajectory and its framing as a 'green city' of the future. Indeed, there are many celebratory parts of the Garden City that have allowed for high standards of living for most, but it is nonetheless important to recognize the other experiences of Singapore that have been – and often still are – left out of the Singapore Story. These experiences include those of LGBTQ+ and Indigenous communities of Singapore, and those of oft-exploited low-wage foreign workers in construction, service and domestic work industries. They also include the many rivers, swamps, mangrove forests, coral reefs and other ecosystems of the islands and the organisms therein that have had their agency and their evolutionary unfolding micromanaged and curtailed by the government.

In her bestselling book *This Is What Inequality Looks Like*, sociologist Dr. Teo You Yenn observes that, 'We [Singaporeans from different socioeconomic classes] come face to face with similar institutions, but we exercise different levels and forms of agency in our interactions with them' (Teo 2019, 113). That is, individual experiences of the structures that the government has put in place to transform Singapore from 'fishing village to a metropolitan city' vary by one's social positions, even when these structures are meant to be utilitarian and beneficial to all. For one, the workers who build and maintain the infrastructure that defines the City in Nature (think construction workers, cleaners, landscapers who prune our trees and cut our grass and more) must endure long hours in the sweltering heat only to be compensated with low wages. Many of these workers are transient migrant workers, whose plight was spotlighted during the COVID-19 pandemic where their cramped and poorly ventilated living quarters led to migrant workers making up ninety per cent of COVID-19 cases at one point, despite only making up five per cent of the population (Lok 2025).

Not all non-humans in the City in Nature are treated equally either – charismatic animals like smooth-coated otters and oriental pied hornbills have become posterchildren for the green city, while others are met with far less warmth. House crows and rock pigeons, for example, are often deemed nuisances and are thus culled (Ganesan 2024; Ting 2023). Not all Nature gets to persist and flourish in the City in Nature, but a ground-up collective called Urban Birds Initiative Singapore is proposing an alternative narrative. The collective is 'striving for a compassionate Singapore where humans, pigeons, crows and mynas can coexist happily without culling' and does so by publishing accessible educational material about these birds on social media and hosting community engagement events ('Urban Birds Initiative SG', n.d.). Similarly, in the remainder of this book, the authors hope to highlight various other narratives that we believe are as important and essential to the Singapore Story as the aforementioned three dominant narratives.

The stories and narratives we will highlight are ones that many international readers are not aware of, unless they make an effort to seek them out or are historians/experts of Southeast Asian politics.[20] While those involved with this book project recognize that 'there exists no conclusive interpretation or definitive consensus on Singapore history' (Sa'at, Joraimi and Sai 2021, 15), all agree that there are some histories that do need (re)telling. All agree that the state-centralized story consistently glosses over (or, per Stibbe and ecolinguistics: erases) key parts of Singapore's history, and given the aims of this book, these key parts will be detailed and brought to the fore in the upcoming sections.

That said, the authors regret that there are still multiple key events, tensions and alternative stories that we were unable to delve deeply into in this book, though they are alluded to and briefly mentioned in various places. This includes the experiences of the LGBTQ+ community, migrant workers, low-income and single-parent households, elderly people, disabled people and the narratives surrounding waste, education, housing, mobility and more. We strongly recommend that interested readers explore the following resources (a non-exhaustive list) to learn more about some of these aforementioned experiences and narratives:

- *Eating Chilli Crab in the Anthropocene: Perspectives on Life in Singapore Perspectives on Life in Singapore* (2020) edited by Matthew-Schneider-Mayerson,
- *Making Kin: Ecofeminist Essays from Singapore* (2021) edited by Andrea Yew, Esther Vincent and Angelia Poon,
- *Migrant Workers in Singapore: Lives and Labour in a Transient Migration Regime* (2022) edited by Brenda S. A. Yeoh, Kellynn Wee and Theodora Lam,
- *Mobilizing Gay Singapore: Rights and Resistance in an Authoritarian State* (2014) by Lynette Chua,
- *Not Without Us: Perspectives on Disability and Inclusion in Singapore* (2023) edited by Kuansong Victor Zhuang, Meng Ee Wong and Dan Goodley, and
- *This Is What Inequality Looks Like* (2019) by Teo You Yenn.

Figure 1.1 Marina Bay.

Figure 1.2 View of Jurong Island and the West Coast of Singapore from Kent Vale Apartments, NUS.

Sustainability in its broadest and most holistic sense requires that these experiences and the human, non-human and overall ecosystems involved with them have to be foregrounded in Singapore's efforts for sustainability in this century of increasing environmental pressures and climate change (Bawaka Country et al. 2015). The reasons behind this claim of centring the other stories that equally make up Singapore will be adumbrated in the articulation of this book's ecosophy in Chapter 2. These other Singapore stories have helped to inform and inspire the authors and our respective case studies in Chapters 3–5, as well (Figures 1.1 and 1.2).

Notes

1 Number as of December 2022 from Department of Statistics Singapore (2025). Note that due to ongoing land reclamation efforts, Singapore's land mass increases year on year.

2 We add to this that Singapore's geographic location and conditions have always played a huge role in Singapore's role in the region and charted its fate as an island. Singapore is located at the tip of the Malay peninsula, surrounded by Indonesia, Malaysia and the straits of Singapore – in fact, one of the mainland's earliest known names was Pulau Ujong. This made it a strategic location for trade, both past (evidence of interregional trade dates back at least 1,000 years) and present.

3 For an overview of this history, see Pwee (2021, accessed 31 January 2023), as well as later chapters in this book.
4 The growth of its infrastructure has established the economic and geographical importance of Singapore's port, which is still a defining feature of Singapore – the ports dominate the skyline along the southern waterfront, demand large-scale heavy infrastructure, and employ thousands. According to the Maritime and Port Authority of Singapore, at the time of writing, the ports of Singapore are linked to 600 ports located in 120 countries around the world, some of which are powerful global forces. Their numbers also suggest Singapore has been one of the world's busiest ports since the 1980s, and that over 130,000 cargo ships make port per annum. Singapore's geographic location and secure maritime location make it the world's leading transhipment hub, where ships dock and transit goods destined for other ports, and load up on cargo for transportation back to the cargo ship's point of origin port. Besides a new container cargo ship arriving every two to three minutes, there are also numerous tourist cruise ships that make port of entry, and then regional ferries that transport passengers to the islands of Singapore, Indonesia, Malaysia and elsewhere. In total, the waterways of Singapore are in a constant state of use and movement, while bringing prestige and economic leverage to the country.
5 Kampung (also spelled 'kampong') refers to the traditional housing settlement, typically stilted houses in close proximity to each other.
6 Approximately 8,300 per square kilometre at the time of writing (or approximately 21,500 per square mile).
7 This will be discussed in later chapters, especially Chapter 5.
8 We recognize that Singapore's port activity, coupled with Singapore's location in Southeast Asia and along the Straits of Malacca, makes Singapore a key economic node in the global economy. This has brought accelerated wealth to the country and a variety of both high-tech and low-tech international and regional labour forces that work in various industries. Such wealth shapes Singapore's growing banking and finance industry and continues to attract investment from abroad, thus fueling the financial industry. These also generate wealth for the government. Singapore's government proactively invests back into the island by building housing for citizens, creating parks and investing in the country's education system, where Singapore's schools are considered some of the best in the world and their public universities consistently rank highly globally. However, layered into this wealth and strategic location of Singapore's port is petroleum (discussed further in Chapter 3). Singapore has one of Asia's largest petroleum refining industries, with Exxonmobil, Shell, Singapore Refining Company, Chevron, Horizon, Meifu, Dulub, NRGLab, China National Petroleum and Apex Development, amongst hundreds of others, all having a presence in Singapore.

These companies refine petroleum into various petrochemical products, including crude oil and store and ship out such products. A vast majority of this refining and distribution occurs on the reclaimed Jurong Island, making this island a top ten world hub by chemical exports volume and a top five refinery export hub. As such, Singapore has a disproportionate impact on climate change given its small geographic size and low population numbers compared to other countries; its per capita energy consumption is amongst the highest in the world, and its prominence as a shipping and refinery hub makes it a central node in the petroleum-based global economy that is driving climate change. And it may be that this reality is part of what drives Singapore's efforts to present itself as a 'green' sustainable city of the future, as for example transitioning Jurong Island into a 'Sustainable Energy and Chemicals Park' under the Singapore Green Plan 2030. Here, the government recognizes that pure petroculture (Wilson, Carlson and Szeman 2017) is not sustainable.

9 It is extremely important to acknowledge both the harm and good that Singapore's particular authoritarian environmentalism has borne – speed of development by most metrics has been accelerated in unparalleled ways, leading to the high quality of life most Singaporeans enjoy. However, this required (and still requires) a certain level of dispossession of peoples and natures.

10 See Civil Service College (2011).

11 https://overshoot.footprintnetwork.org/newsroom/country-overshoot-days/ Accessed 30 May 2025. That is, if everyone on Earth lived like Singapore, the Earth's annual resources would be depleted before a quarter of the year concluded.

12 Lee Kuan Yew is widely considered one of the first founding fathers of the country and has been given the status of the 'Architect of Modern Singapore' (Sim 2015). His vision is recognized to have played a pivotal role in building the Singapore of today.

13 For an overview of this history, see National Library Board Singapore (2015), accessed 27 January 2023. The Garden City narrative will be further analysed in Chapter 4's case study on the City in Nature discourse.

14 Note that Prime Minister Lee ended his term in 2024 and he is the eldest son of Lee Kuan Yew.

15 All quotes from Prime Minister's Office Newsroom coverage of speech (2012, accessed 28 January 2023).

16 ASEAN stands for the Association of Southeast Asian Nations.

17 This discourse is further analysed in Chapter 4's case study on City in Nature.

18 More on this will be analysed in Chapter 4's City in Nature section and Chapter 5's Park Connector Networks section.

19 See Chapter 3's section on Pulau Bukom for more on Singapore's 'oil islands'.

20 Note from the editor: As an anecdote of one located within such an international context, I (Todd) place myself in this category. This is because

before relocating to Singapore, I had only encountered the dominant 'Singapore Story', in both its form of government and that government's development; and in its packaging as a sustainable city of the future. I am thankful for my collaborators in this book project, for their narration of the history of this other story's formation and its ongoing telling and their articulation of why knowing this other story is essential to understanding Singapore, and thus sustainability in Singapore. For sympathetic yet strategic insights from a foreigner who married a Singaporean and who worked in Singapore, especially about the various social, public health, political and technological transitions of the 1950s–1970s, see Wee (2017).

References

Afandi, Suhaimi, and Ivy Maria Lim. 2022. 'History Education in Singapore: Development and Transformation'. In *Education in Singapore*, 379–403. Singapore: Springer Singapore. https://doi.org/10.1007/978-981-16-9982-5_21.

Barr, Michael. 2014. *The Ruling Elite of Singapore: Networks of Power and Influence*. London: I.B. Tauris.

Chua, Beng Huat. 2009. 'Being Chinese Under Official Multiculturalism in Singapore'. *Asian Ethnicity* 10 (3): 239–50.

Chua, Beng Huat. 2017. *Liberalism Disavowed: Communitarianism and State Capitalism in Singapore*. Singapore: National University of Singapore Press.

Chua, Ee Kiam. 2015. *Rainforest in a City*. Singapore: NParks.

Chua, Lynette J. 2014. *Mobilizing Gay Singapore: Rights and Resistance in an Authoritarian State*. https://www.academia.edu/11331562/Mobilizing_Gay_Singapore_Rights_and_Resistance_in_an_Authoritarian_State.

Civil Service College. 2011. 'Developing Our Approach to Public Engagement'. Accessed October 2011. https://knowledge.csc.gov.sg/ethos-issue-10/developing-our-approach-to-public-engagement/.

Connolly, Creighton and Hamzah Muzaini. 2022. 'Urbanizing Islands: A Critical History of Singapore's Offshore Islands'. *EPE: Nature and Space* 5 (4): 2172–92.

Country, Bawaka, Sarah Wright, Sandie Suchet-Pearson, Kate Lloyd, Laklak Burarrwanga, Ritjilili Ganambarr, Merrkiyawuy Ganambarr-Stubbs, Banbapuy Ganambarr, Djawundil Maymuru and Jill Sweeney. 2015. 'Co-Becoming Bawaka'. *Progress in Human Geography* 40 (4): 455–75. https://doi.org/10.1177/0309132515589437.

Department of Statistics Singapore. 2025. Environment. Accessed 1 January 2025. http://www.singstat.gov.sg/publications/reference/ebook/society/environment.

Er, Kenneth. 2020. 'Transforming Singapore into a City in Nature'. *Urban Solutions*, 19: 68–77.

Fujii, Tomoki, and Ray Rohan. 2019. 'Singapore as a Sustainable City: Past, Present and the Future'. *SSRN Electronic Journal*. https://doi.org/10.2139/ssrn.3480894.

Ganesan, Natasha. 2024. 'Pigeon Culling Among Measures to Be Stepped up in Pilot Initiative at 3 Town Councils'. *CNA*, 26 June 2024. https://www.channelnewsasia.com/singapore/ang-mo-kio-bishan-toa-payoh-tanjong-pagar-town-councils-pigeon-culling-nparks-4434341.

George, Cherian. 2010. *Singapore: The Air-Conditioned Nation-Essays on the Politics of Comfort and Control 1990–2000*. Singapore: Landmark Books.

Ghosh, Amitav. 2021. *The Nutmeg's Curse: Parables for a Planet in Crisis*. Chicago: University of Chicago Press.

Gilley, Bruce. 2012. 'Authoritarian Environmentalism and China's Response to Climate Change'. *Environmental Politics* 21 (2): 287–307. https://doi.org/10.1080/09644016.2012.651904.

Global Footprint Network. 2025. 'Ecological Footprint'. Global Footprint Network. https://www.footprintnetwork.org/our-work/ecological-footprint/.

Goldstein, Jesse and David Tyfield. 2017. 'Green Keynesianism: Bringing the Entrepreneurial State Back in(to Question)?' *Science as Culture* 271 (1): 74–97.

Guo, Remy. 2016. *Urban Redevelopment: From Urban Squalor to Global City*. First edition. Urban Systems Studies. Singapore: Centre for Liveable Cities.

Han, Heejin. 2017. 'Singapore, a Garden City: Authoritarian Environmentalism in a Developmental State'. *The Journal of Environment and Development* 26 (1): 3–24. https://doi.org/10.1177/1070496516677365.

Hobson, Kersty. 2005. 'Considering 'Green' Practices: NGOs and Singapore's Emergent Environmental-Political Landscape'. *Sojourn: Journal of Social Issues in Southeast Asia* 20 (2): 155–76.

Housing Development Board. 2024. 'Public Housing – A Singapore Icon'. Accessed 17 April 2024. https://www.hdb.gov.sg/about-us/our-role/public-housing-a-singapore-icon.

Kong, Lily. 2013. 'Tan Chuan-Jin: Finding Ways to Collabourate'. Interviewed by Lily Kong. *Centre for Liveable Cities*, 7 May. https://www.clc.gov.sg/docs/default-source/urban-solutions/urb-sol-iss-3-pdfs/interview-tan-chuan-jin.pdf.

Kuo, Eddie C. Y., and Brenda Chan. 2016. *Singapore Chronicles – Language*. Singapore: Institute of Policy Studies: Straits Times Press.

Lee, Hsien Loong. 2012. 'PM Lee Hsien Loong at the Opening of Gardens by the Bay'. Text. Prime Minister's Office Singapore. 28 June 2012. https://www.pmo.gov.sg/Newsroom/speech-prime-minister-lee-hsien-loong-opening-gardens-bay

Lim, Rachel. 2025a. '43 Workplace Deaths in Singapore in 2024, Up From Previous Year'. *CNA*, 28 March 2025. https://www.channelnewsasia.com/singapore/workplace-deaths-major-injuries-mom-construction-sector-5024886.

Lim, Yan Liang. 2025b. 'GE2025: PAP Gets 65.57 per Cent of Votes in Landslide Win'. *The Straits Times*, 4 May 2025. https://www.straitstimes.com/singapore/politics/ge2025-pap-gets-65-57-per-cent-of-votes-in-landslide-win.

Liow, Eugene Dili. 2012. 'The Neoliberal-Developmental State: Singapore as Case Study'. *Critical Sociology* 38 (2): 241–64. https://doi.org/10.1177/0896920511419900.

Lok, Cheryl. 2025. 'Singapore's "Invisible" Population: The Perpetual "Homelessness" of Migrant Workers'. *Oxford Political Review*, 25 April 2025. https://oxfordpoliticalreview.com/2025/04/25/singapores-invisible-population-the-perpetual-homelessness-of-migrant-workers/.

Low, Linda. 2001. 'The Singapore Developmental State in the New Economy and Polity'. *The Pacific Review* 14 (3): 411–41. https://doi.org/10.1080/09512740110064848.

Millet, Didier, ed. 2013. *Living in a Garden: The Greening of Singapore*. Singapore: NParks.

Ministry of Sustainability and Environment, Ministry of Education, Ministry of National Development, Ministry of Transport, and Ministry of Trade and Industry. 2021. 'Singapore Green Plan 2030'. *Singapore Green Plan 2030*. Accessed February 2021. https://www.greenplan.gov.sg/overview/.

National Library Board Singapore. 2015. '"Garden City" Vision Is Introduced'. March 2015. https://www.nlb.gov.sg/main/article-detail?cmsuuid=a7fac49f-9c96-4030-8709-ce160c58d15c.

Plumwood, Val. 2008. 'Shadow Places and the Politics of Dwelling'. *Australian Humanities Review* 44. https://australianhumanitiesreview.org/2008/03/01/shadow-places-and-the-politics-of-dwelling/.

Ponte, Stefano. 2023. 'Green Capitalism and Unjust Sustainabilities'. *Samfundsøkonomen* 2019 (4): 102–8.

Public Utilities Board. 2023. 'Singapore's First Centre of Excellence to Drive Coastal Protection and Flood Management Research'. PUB, Singapore's National Water Agency. Accessed 7 September 2023. http://www.pub.gov.sg/Resources/News-Room/PressReleases/2023/09/Centre-of-Excellence-to-drive-coastal-protection-and-flood-management-research.

Pwee, Timothy. 2021. 'From Gambier to Pepper: Plantation Agriculture in Singapore'. Accessed 1 April 2021. https://biblioasia.nlb.gov.sg/vol-17/issue-1/apr-jun-2021/agriculture/.

Rowe, Peter G., and Limin Hee. 2019. *A City in Blue and Green: The Singapore Story*. Singapore: Springer Singapore. https://doi.org/10.1007/978-981-13-9597-0.

Sa'at, Alfian, Faris Joraimi and Siew Min Sai. 2021. 'Introduction'. In *Raffles Renounced: Towards a Merdeka History*, 11–16. Singapore: Ethos Books.

Schneider-Mayerson, Matthew. 2020. 'Introduction: Seeing Singapore with New Eyes'. In *Eating Chilli Crab in the Anthropocene*, 9–18. Singapore: Ethos Books.

Siddiqui, Kalim. 2016. 'A Study of Singapore as a Developmental State'. In *Chinese Global Production Networks in ASEAN*, 157–88. Understanding China. Cham: Springer International Publishing. https://doi.org/10.1007/978-3-319-24232-3_9.

Sim, Walter. 2015. 'Mr Lee Kuan Yew Was Architect of Our Modern Republic, Says President Tan'. *The Straits Times*. https://www.straitstimes.com/singapore/mr-lee-kuan-yew-was-architect-of-our-modern-republic-says-president-tan.

Singapore's Voluntary National Review. n.d. https://www.mfa.gov.sg/SINGAPORES-FOREIGN-POLICY/International-Issues/Singapore-Voluntary-National-Review.

Tabb, Phillip. 2021. *Biophilic Urbanism: Designing Resilient Communities for the Future*. New York: Routledge.

Tan, Puay Yok, Wang, James and Sia, Angelia. 2013. 'Perspectives on Five Decades of the Urban Greening of Singapore'. *Cities* 32: 24–32. https://doi.org/10.1016/j.cit ies.2013.02.001.

Teo, You Yenn. 2019. *This Is What Inequality Looks Like*. Singapore: Ethos Books.

Tham, Aaron. 2019. 'Envisioning Eden: The Manufactured Ecotourism Environment of Singapore'. *Journal of Ecotourism* 18 (1): 1–24. https://doi.org/10.1080/14724 049.2017.1334787.

Ting, Koh Wan. 2023. 'Better Ways to Control Crow Numbers Than Culling Them, Says ACRES After Bishan Crow Attacks'. *CNA*, 22 February 2023. https://www.channeln ewsasia.com/singapore/crow-culling-better-ways-control-acres-nparks-bishan-att ack-3291266.

Turnbull, C. M. 2009. *A History of Modern Singapore, 1819–2005*. Singapore: NUS Press.

Wee, Ann. 2017. *A Tiger Remembers: The Way We Were in Singapore*. Singapore: NUS Press.

Wilson, Sheena, Adam Carlson and Imre Szeman, eds. 2017. *Petrocultures: Oil, Politics, Culture*. Chicago: McGill-Queen's University Press.

Yeoh, Brenda S. A., Kellynn Wee and TheodoraLam, eds. 2022. *Migrant Workers in Singapore: Lives and Labour in a Transient Migration Regime*. Singapore: World Scientific. https://doi.org/10.1142/12798.

Yew, Andrea, Esther Vincent, Angelia Poon and Ethos Books, eds. 2021. *Making Kin: Ecofeminist Essays from Singapore*. Singapore: Ethos Books.

Yiannouka, Stavros N. 2015. 'The Secret of Singapore's Success in Education'. *The Straits Times*, 11 April 2015. https://www.straitstimes.com/opinion/the-secret-of-singapo res-success-in-education.

Zhuang, Kuansong Victor, Meng Ee Wong and Dan Goodley, eds. 2023. *Not Without Us: Perspectives on Disability and Inclusion in Singapore*. Singapore: Ethos Books.

2

Ecosophy

Todd LeVasseur, Kristina Gweneth Ponce Simundo and Lecarl Lim

Journeying from Ecolinguistics to Environmental Philosophy and Ethics (and Back Again)

A key insight of ecolinguistics is that the language used by humans in the *stories-we-live-by* shapes our perception, and thus in part our treatment of, the more-than-human world. That grammar, discourses and semiotics – the texts that we engage, their construction and how these texts inform our symbolic consciousness and worldviews – will nudge or inform our behaviours in relation to nature or the environment at subconscious and unconscious levels. As Arran Stibbe writes, 'In essence, ecolinguistics consists of questioning the stories that underpin our current unsustainable civilisation, exposing those stories that are clearly not working, that are leading to ecological destruction and social injustice, and finding new stories that work better in the conditions of the world that we face' (Stibbe 2014, 117).

Here we note that ecolinguistics, at least as argued for by Stibbe, is normative in its methods and deployment. This means there is a recognition that dominant industrial and extractive-based lifeways are currently unsustainable, both socially and environmentally, and thus they need to change (see also the introduction for more on this). Moreso, this change will therefore require (i.e. normatively necessitates) new *stories to live by*, and thus new stories produced and encoded in forms of grammar and in texts. We are in agreement that ecolinguistics is and must therefore be normative. We argue that ecolinguistics must also be informed by the feminist tradition of pedagogy (hooks 2014), and we further hold that any and every discipline[1] in higher education, including ecolinguistics, must also be normative in regards to social and environmental issues.

How, though, are we to gauge the efficacy of new stories – what should they espouse, celebrate, inform and inspire in their respective audience/s, and why; and how would we know if they are successful? What metrics can and should be used, to both critique existing and often hegemonic stories and their destructive ideologies, that would therefore allow us to criticize them as ideologically unhealthy in relation to the environment (and to social metrics)? These are important normative questions to answer. They lead us to another key question: what standards may we bring to bear as we engage with and analyse dominant discourses, and how those dominant discourses may impact social and environmental metrics? At their core, these are questions of relationality, both amongst humans, and between humans and other biotic and abiotic aspects of our planet. They are also methodological and theoretical questions of measurement based on some measure of using standards of social and environmental health and flourishing in analysing discourses. They are also questions of intent and purposeful creation by those making the *stories we live by*. They are thus questions about how those stories encode (or challenge) existing worldviews, while encoding the worldview of the creator/s of the discourse.

Relationality, metrics of health, worldview and values – these are concepts and categories central to ethics. Here, ethics refers to the norms, obligations/duties and concepts of good or bad or virtuous ways of living that are used to judge and reflect upon human actions and behaviours as these occur in a community setting. Or, as the energy humanities scholar Imre Szeman explains, 'Whatever else ethics might be, they are intended to constitute the elabouration of the axioms and principles around which action and practice is shaped and governed, with respect to individuals, groups, or even non-human beings' (Szeman 2013, 158).[2]

Consistent with the methodology of Stibbe (2014), in this chapter, we will create[3] an ecosophy that contains axioms and principles around which everyone but Todd will in subsequent Chapters 3–5 judge various self-selected stories, texts and discourses present in Singapore that are either *destructive*, *ambivalent*, or *beneficial*. These are stories and discourses that inform action and practice in the varied ecologies of living places in Singapore, as they are the stories that Singaporeans live by. They are Singaporean stories that encode ideologies and therefore contain ethical elements and normative implications for how Singaporeans should live.

These stories also inform subject formation, and therefore, subjectivity where subjectivity can be understood as the 'modes of perception, affect, thought, desire, fear, and so forth that animate acting subjects[,] as well as the cultural and social formations that shape, organize, and provoke those modes of affect, thought and so on' (Ortner 2005, 31). We also understand subject formation to be an iterative process: our subjectivity is shaped and informed by the stories we live by and the messages, ideologies, framing and other aspects they share and encode, while these stories also help shape and form larger social forces. The latter thus also shape the former. And both shape and impact the ecologies of living places within which we always act out our subjectivities and enact who we are as humans in biosocial communities.

Our ecosophy, therefore, will allow us to interrogate, analyse, deconstruct and evaluate the ethics of relation and representation towards the more-than-human world as contained in the environmentally focused (or absenced) stories in Singapore that have been curated in Chapters 3–5 and that inform Singaporean subjectivities. Before the ecosophy that motivates the book and how it was formed is shared, however, we need to first share a brief history of environmental philosophy (also environmental ethics). Note that we focus on its development in the West/Global North, to situate what an ecosophy is and how it may work, as it is within Western/Global North environmental philosophy that the specific concept of an ecosophy as developed by Arne Naess (as well as the field of ecolinguistics) emerged.[4]

Given the above, we recognize there can be, and are, Asian and indigenous environmental philosophies, including those found in Singapore. Many of these relate to human–human relations that are thoroughly influenced and shaped by Chinese conceptions of filial piety and duty, or Buddhist conceptions of *karma* and compassion and rebirth,[5] or Hindu conceptions of *dharma* (duty), while many of the ethics in these systems are gendered and related to birth order and age. However, to date, dominant discourses about appropriate human–nature interactions, especially in a globalized context, and in the context of sustainable development, have largely emerged out of or been a result of hegemonic Western discourses and hegemonic economic and political power built on colonialism and imperialism. Therefore, they are heavily influenced by Western environmental philosophy and ethics and also environmental policy, where the latter is in part informed by the former.

It is the case that as explained in this book's Chapter 1, per the governmentally created 'Singapore Story', that Singapore as a governed country does largely understand nature in a way as a collective resource that should benefit everyone. As explained there, this story is also informed in large part by the PAP's conception of a strong, supportive but firm bureaucratic ruling parent in the form of the government. Therefore, this meta-level governmental view, informed as it is by conceptions of filial duty and care, of 'liberalism disavowed'

(Huat 2017), still occurs within an instrumental ethical story and paradigm of seeing nature as a resource. This collective subjectivity is consistent with classical and neoliberal economics that emerged in Europe (Moore 2017) and that have been codified in globalization via Washington and post-Washington Consensus politics (Sheppard and Leitner 2010; Ban and Blyth 2013) and a wide variety of free trade agreements historically based upon such politics.

Given the above, then, we want to be clear that in the creation of this book's ecosophy, this creation is being done within a Western method as articulated and modelled by Stibbe. However, the authors other than Todd will be weaving into this book's ecosophy standards, axioms and ideals that are very much shaped by the shared lived experience of living in Singapore and of being in the ecologies of alive places that make Singapore and that therefore actively influence their subjectivities. It is indeed the case, and as we will see below when we discuss the function of an ecosophy, that an ecosophy must be attuned to the ecology, environment and human systems of a place. This is because these shape the values, ethics and stories of earthen-care for the people in those places who labour to develop and live by an ecosophy. Therefore, this immediate section of this chapter will focus much more on Western environmental philosophy and ethics and this intellectual tradition, while later sections of the chapter written by authors other than Todd will include Asian and indigenous environmental philosophy, as well as understandings of human–human and human–nature relations that are very much shaped by the lived experience of being in Singapore.

Western Environmental Philosophy

Western environmental ethics and philosophy, as both a real and imagined tradition, often begins with the work of the United States (US) forester and conservationist, Aldo Leopold, and his posthumous 1949 collection of essays, *A Sand County Almanac*. In this book, Leopold argued for the development of a 'land ethic', where human actions are measured by their ability to either perpetuate or create beauty in the biotic community of life or by how human actions may destroy this community of life (1949). Via his essays, Leopold's environmental ethic shifted the locus of concern from an anthropocentric, human-centred metric to a biocentric, or land-centred metric. Following this, in 1971, the philosopher J. Baird Callicott, inspired by both Leopold and the historian Lynn White, Jr.'s 1967 article 'The Historical Roots of our Ecological Crisis', taught the first environmental philosophy and ethics course in the history of higher education at the University of Wisconsin-Stevens Point (Callicott 2017). Very quickly, other philosophers turned their training on deontological questions of rights, duties, rules and obligations, theories of virtue and explorations of value systems to analyse and investigate the ethics of human–environment interactions. These questions in the late 1970s and early 1980s, driven in part by the academic platform of the journal *Environmental Ethics* that formed in 1979, also quickly turned to animal rights and also issues related to the conservation of nature and the rights of nature.

In 1981, another early and now prominent environmental ethicist and philosopher, Holmes Rolston, III, systematically evaluated 'nature' as a category and placeholder of value, and thus the various ways humans may value nature. His list includes economic value, life support value, recreational value, scientific value, aesthetic value, life value, diversity and unity values, stability and spontaneity values, dialectical value and sacramental value (Rolston 1981). To this, we add values of refuge and escape (especially from oppression, as in maroonage), solace, health and well-being, humour and sexecology or the erotic. We also note there are 'bad environmentalisms' (Seymour 2018), and that values of and invested into nature may occlude or downplay the very real harms humans may and do suffer from natural disasters, zoonotic diseases, pathogens, viruses, extinction of other species and invasive species.[6]

A variety of approaches emerged besides those mentioned above as Western environmental philosophy continued to develop into a systematic and interdisciplinary field of theorizing environmental ethics. One of these approaches is the social ecology of the US social theorist and political anarchist, Murray Bookchin, and his philosophy of ecosocialism or ecological socialism.[7] Broadly speaking, though, today his work is not as systematically engaged and does not have the audience it did in the 1980s and 1990s. Another approach was put into practice with the formation in the 1980s of radical environmentalism, as both philosophically and tactically seen in groups like Earth First!, the Earth Liberation Front and the Animal Liberation Front (Taylor 1998). These groups centre an ecocentric (i.e. ecosystem and biotic and abiotic attributes of) worldview and feel obligated to engage in tactics such as property destruction in defence of wild nature and non-human organisms. Their views were/are seen by some critics as misanthropic, and their actions have even received charges of ecoterrorism (Taylor and LeVasseur 2008) by some governments. The radical environmentalist approach to wilderness preservation and sustainable living helped to create the bioregional movement, another philosophical approach to living within the biogeochemical identity of, and ecoscape flows found within, watersheds and other ecosystem boundaries (LeVasseur 2011).

A strident critique of philosophy, broadly, also emerged from the rise of first- and second-wave feminisms. A key criticism that emerged during this time challenged mainstream Western philosophy as being entirely androcentric and limited by studying a largely European male canon. Another feminist critique of philosophy as it was conceived, taught and practiced at this time was that only rational approaches to systematic philosophy were seen as valid in mainstream philosophy. In response, feminist philosophers examined ethics from and through what became known as the feminist care tradition, where relationality, especially based on feelings and effects of care, trust and love form the basis of how, when and why humans value 'others' in their relations.[8]

Within environmental philosophy/ethics, proper, this feminist critique of philosophy joined with rising concern about ecological metrics in the late 1960s, 1970s and 1980s to become ecofeminism. Ecofeminists began to argue that hierarchical logics of domination based on policed binaries, such as the exclusionary binaries of male–female, human–nature, objective–subjective, human–animal and rational–emotional, where the first terms and categories are ontologically and epistemologically superior to the latter terms and categories, form interlocking oppressions and systematic degradation of both female agency and bodies. Moreso, they combine to also systematically degrade, oppress and exploit the earth, too (Warren 1990; Plumwood 1986; Cheney 1987). Ecofeminism and its focus on the dual and interlinked oppressions of females and nature helped to inform the subsequent formation of queer ecologies that recognizes the analysis of early ecofeminism was too often bioessentialist and also trans-exclusionary (Sandilands 2016). Ecofeminism also informed the formation of ecowomanism (Harris 2016) which holds similar ethical analyses as ecofeminism but foregrounds in its project how certain bodies are also racialized, and this trifecta of race/gender/ecology needs to be foregrounded in ethical analyses. A parallel development from front-line activists and scholars of colour centred the concepts of environmental justice and environmental racism (Bullard 1993) with calls for distributive, restorative, participatory and representational justices and practices. Ecological masculinities is another environmental philosophy and normative approach to analysing human–nature interactions that has recently emerged in this lineage, with a specific focus on how masculinities shape relations to self and others, where the 'other' includes the more-than-human (Hultman and Pulé 2018).

More recently, in environmental philosophy and ethics, there is a recognition of indigenous philosophies and calls to decolonize ethics, broadly, and to de-centre settler ecologies (Whyte 2013, 2018). Meanwhile, other environmental philosophers have begun to argue for a more nuanced analysis of relationality (vs. either intrinsic or instrumental approaches to valuation) as an entry into concepts of value, ethical judgements and appropriate behaviours (Stalhammar and Thoren 2019). Other ways in which environmental philosophy and ethics continue to develop and branch out, building on the initial heritage of Leopold's land ethic, now includes asking what does environmental philosophy even mean on a planet without ice caps (LeVasseur 2014), to expanding the concept of species-being to non-human organisms (Fair and McMullen 2023), to arguing for a 'just transition' to a post-carbon future that takes climate justice as a central charge (Wang and Lo 2021), to exploring questioning the very category of 'nature' given the reality of hyperobjects (Morton 2012, 2013).

Astute readers familiar with environmental philosophy and ethics will note a glaring gap in this overly quick summary[9] of Western environmental philosophy: that of deep ecology. We now turn our attention to providing an overview of this important development within environmental ethics. We provide deep ecology with its own sustained focus for the concept and evaluative project of an ecosophy that Stibbe employs and that we employ in the ecolinguistical methodology of this book, which emerges from this branch of environmental philosophy and ethics.

Deep Ecology and Its Praxis

Deep ecology as a concept and term originated with the Norwegian philosopher, Arne Naess (1912–2009). Besides being Norway's best-known philosopher, with his work influenced in part by Spinoza (Naess 1977), Naess was also an avid hiker and mountaineer. Naess's political and ecological thought was also informed by Gandhi's non-violence and *satyagraha* (spiritual pathway to embodying one's truth in the face of oppression). These influences coalesced when Naess famously criticized the mainstream Western environmental groups of the 1970s as being 'shallow' in their critique of root environmental causes, as being too bureaucratic and as being too willing to compromise. Rather, Naess argued the need for a 'deep' ecology: to deeply question the guiding norms, values and perceptions of the natural world that he argued led to the ecological crisis (Naess 1973, 1986b). Central to Naess's thesis and argument is that mainstream, shallow ecology is at its core instrumental and anthropocentric in its view of nature; that it has not taken the insights of ecology seriously enough, including the insight that everything is interconnected and humans are therefore not unique, special, or separate from the rest of life. Compared to this, a deep ecological view is based on an ecocentric ethical bearing that sees all life, and thus all life forms, as having intrinsic value. Furthermore, deep ecology holds that all life forms have the right to flourish and live out its/their evolved species-being, with a concomitant decentring of the human animal from its assumed supremacy and its instrumental bearing towards human–nature relations.

Naess's work quickly gained followers and interlocutors and also gained various critiques. In terms of criticisms, key ones emerged from the Global South, and especially from Ramachandra Guha, who argued that deep ecology's privileging of ecology and discussion of population size could be read as being anti-human and imperial. He further argued that the lack of sustained discussion of economic and political disparities by deep ecologists meant that proponents of deep ecology, and the promotion of deep ecology, were not relevant to victims of globalization and who lived in extreme poverty. Lastly, he opined that the true causes of environmental destruction were unchecked consumption and militarization, and solving these required targeted and structural policy interventions, not the recognition of intrinsic value of all organisms (Guha 1989).[10] We are sympathetic with the critiques of Guha, especially as outside of Todd, we are located as authors in an over-consuming country surrounded by neighbouring countries where people experience abject levels of poverty, suffering and hardship. However, we do note that the move to ecocentrism and the development of an ecosophy as an ecolinguistical methodology are both relevant and helpful, especially in criticizing the very same over consumption about which Guha wrote, including in Singapore.

Other critiques of deep ecology are also worth briefly mentioning and taking seriously. On the difficulties in moving beyond anthropocentrism broadly, even in deep ecology, see Katz (2000); and on the import of ecocentrism in nature conservation, a position we hold given Singapore's high material standard of living and precarious island geography, see Washington et al. (2017). Further critiques emerged from Richard Sylvan, who argued that at its core Naess's call for a biocentric egalitarianism is untenable and leads to inconsistencies, and moreso, it is not required that ecological self-realization is a value that people are obligated to maximize (1985).[11] Further criticisms that emerged over the years include charges that deep ecology is a functional equivalent of a religion and may therefore slide into fundamentalism (Devall 1996); that the quest for an ecological 'Self' may lead to forms of gendered tyranny as well as human exceptionalism (Plumwood 1991, 1999); and that deep ecology lacks clarity in how humans will have to navigate inevitable competition in people's values and in their ultimate premises (Glasser 2011).[12] Taken together, these are all important critiques of deep ecology that we encourage readers to engage and upon which they may want to reflect. Yet, we recognize any and all discourses, let alone an entire environmental philosophy, will always have criticisms. There is no 'perfect' discourse to guide human and more-than-human interactions in the Anthropocene, let alone a perfect environmental philosophy.

Naess and other deep ecologists were of course cognisant of these and similar criticisms as such emerged over the decades. Some of the earlier criticisms Naess attempted to resolve while at the same time more systematically developing deep ecology as a philosophical approach. He concurrently attempted to move deep ecology into a basis of ecological praxis. This move by Naess was aided in his relationship with the US environmental philosopher George Sessions. During a camping trip in 1984, they developed what became the 'Eight point platform' of deep ecology (Table 2.1).

Note that this platform is not prescriptive in terms of obligating prescribed behaviours and duties. Rather, as Sessions shares, 'Naess describes ecological egalitarianism [the implications of premises one and two] as an intuition experienced by those in the deep ecology movement, not an ethical theory to be defended by rational argument' (Sessions 1987, 117). Key, here, is that one must experience and intuit and reflect upon, at a deep ontological level, one's embeddedness in a fecund, living, intrinsically valuable and biologically rich world of life. The hope is that as one experiences this world, and sees and knows its value and richness, it will inform and inspire a shift from the shallow ego 'I' (and the ideologies of consumption, instrumental valuing of nature and human supremacy) to a deeper ecological 'Self'. As Naess explains, 'I shall offer only one simple sentence that resembles a definition of the ecological self: The *ecological self* of a person is that with which this person identifies. The key sentence (rather than a definition) about the self shifts the burden of clarification from the term *self* to that of *identification*, or rather, the process of identification' (1986a, 25–6, italics his). For Naess, this identification is one that moves outwards to include all of life, as a form of ecological egalitarianism and metaphysical holism: when one identifies with life, one will protect it, cherish it, respect it, be inspired by it and want to live sustainably within it (thus inspiring the work implied in premises five to eight).

Table 2.1 The Deep Ecology Platform

One-word summary of the derived value contained in the premise:	**Core values of deep ecology that build upon one another, beginning with core value/premise #1:**
Life!	1. All life has intrinsic value, and all beings have value in themselves, independent of instrumental use value for humans.
Nature!	2. The diversity of life, of nature and ecosystems is itself a value, and this diversity is central to the unfolding of life and the opportunity of all organisms to realize their own intrinsic value.
Humans in/of nature!	3. Humans have no right to lessen this diversity except to satisfy vital needs.
No false distance!	4. Currently, humans are massively interfering with the health and well-being of the natural world, and this is only getting worse.
Outside change!	5. In order to allow for human life to flourish, there needs to be less humans; in order to allow for non-human life to flourish, there must be less humans.
Inside change!	6. All of the above imply that there must be economic, political, technological and other types of changes in how humans live. It also implies a change in ideology, away from anthropocentrism and shallow ecology, towards deep ecology. These changes mean the future will look very different than the present.
Spread of ideas!	7. The ideological change will also centre upon recognizing the inherent worth of life quality, rather than concepts of a good life or happiness as being determined by an endless pursuit of higher standards of living and consumption.
Action!	8. If someone agrees with deep ecology premises #1 through #7, then they are obligated to either directly or indirectly participate in the required and necessary changes referenced in premise #6.

Note this is adapted from David Rothenberg (1987, 186, especially the left column) and also Naess and Sessions (1986).

> Further, for Naess, we need environmental ethics, but when people feel that they unselfishly give up, or even sacrifice, their self-interests to show love for nature, this is probably, in the long run, a treacherous basis for conservation. Through identification, they may come to see that their own interests are served by conservation, through genuine self-love, the love of a widened and deepened self. At this point, the notion of a being's interests furnishes a bridge from self-love to self-realization. (1986a, 29)

This self-realization is based in part upon recognizing, pace the insights of ecology, that 'The "everything hangs together" maxim of ecology applies to the self and its relation to other living beings, ecosystems, the ecosphere, and the earth, with its long history' (1986a, 32).

In many ways, the deep ecology platform, as with any worldview that contains a strong ethical component, becomes a *story to live by*. To wit, the platform anchors and informs an ecocentric identity and contains an ideology that holds that all life has intrinsic value and humans need to stop destroying the rest of life by changing their anthropocentric and shallow ideologies. Furthermore, this claim is based on insights from ecology and evolution. This story also implies a deeper questioning about and reflection upon what constitutes the 'self', which for Naess is really a Self that extends its identity and allegiances outwards to include the rest of life. For Naess, one should ideally realize this through deep reflection and deep questioning, based on self-formed and directed ultimate premises: a short, succinct series of statements, informed by the platform, about how one wants to try to live in a way that recognizes the interconnectedness of all of life. This series of succinct ultimate premises is an ecosophy: from the Greek *oikos*, for house or home, and the Greek *sophia*, for wisdom. An ecosophy, then, is a strategy and approach to cultivating deep ecological wisdom contained in a series of reflective, pithy statements, that is generated by a Self, where this ecosophy is attempted to be put into practice as a story to live by as the creator of the ecosophy navigates their daily life.

We end this section of the chapter, reflecting on this quick journey through Western environmental ethics, with a revisit about why we need an ecosophy. For our ecolinguistical methodology, an ecosophy becomes something through which the *stories-we-live-by* (and the subjectivities they help to form and the behaviours they influence) can be measured, critiqued and critically analysed. The data are clear that we are in a planetary ecological crisis, as explored earlier and throughout this book. That crisis will (and already does) have severe implications for Singapore and Singaporeans and for the sustainability of Singapore. The size, scale, complexity and urgency of these 'environmental issues and their consequences call for important social transformations (acting and thinking differently). Therefore, thinking about environmental ethics should be a fundamental reflection that allows us to establish norms, rights and human duties towards nature' (Valbrun 2023, 1031). Creating an ecosophy, or ecosophies, can allow Singaporeans to reflect on and create norms, rights and human duties that can help lead to *beneficial* discourses and ideally transformative ways of living in Singapore. Moreso, developing an ecosophy grounded within an environmental ethic, or rather, an ecosophy as a living-out-a-practice of an environmental ethic,

> refers to a triple relation: relation of oneself to the other, relation of oneself to the cosmos, relation of oneself to life ... It must necessarily analyze the relations between nature and humans in a reflexive way by considering the proper needs of both while taking into account their impacts in space and in time with regard to moral values. (Valbrun 2023, 1031)

This book's ecosophy reflects on appropriate moral values, behaviours, duties and obligations in regards to these three relations. It also considers individual, human–human and human–nature needs in Singapore within the context of planetary environmental harm and also climate triage and suffering.

Our Group Ecosophy

This book entails what we[13] believe is the first attempt to create a group ecosophy that will be used to undertake an ecolinguistical analysis of a city. Our goal was to speak to the dominant *destructive* discourses we have briefly outlined in the introduction and Chapter 1 and to amalgamate our shared hopes, dreams and visions for Singapore and the world at large into a cohesive vision.

Both the process of forming our ecosophy and the explanation that we present in this chapter are dialectically interconnected – our shared ecosophical principles prioritize a collaborative process, and it is through our dynamic and interpersonal relationships within the group both in the context of the book and outside of it that we have shaped the formation of a shared ecosophy. Furthermore, our ecosophy also upholds the significance of iteration and is wary of the dangers of a static, unchallenged vision for the future. As follows, the ecosophy we present below is the iteration we settled on for this book and as a lens to analyse the case studies we have picked out in Singapore, but we present it with the knowledge that the ideas in it are still evolving. Nonetheless, the following section details our process of forming our ecosophy, a process that is a microcosm of how we hope people with pluralistic views of climate futures can come together to form a shared vision. Following this, we share our ecosophy and elaborate on the principles that underpin it.

Process: Coming Together to Build a Shared Hope for the World

On 25 October 2023, around a small metal coffee table, at the peak of the afternoon heat, the six of us contributing authors began a two-year-long journey together. It was to be a fluctuating journey between the tensions and joys of bringing together a group of humans hoping to create something bigger than the sum of their whole. We had to navigate the ways our individual values and personalities worked together and also differentiated. We had to hold an open, inclusive and equitable space to help each other coax out our principles into coherent and succinct language. We also had to manage our differences in our personal ecosophies – finding compatibility and compromise, while allowing for individuality and plurality. It was no doubt challenging, but a necessary process guided by pluralism as a principle, to transcend the traditional boundaries of academia in order to produce transdisciplinary knowledge and a research praxis (Constanza et al. 2020).

Our first meeting began with us reflecting on our personal experiences, learning and ambitions that have shaped our personal eco-values. Each of us shared commonality in being students from the same B.A. programme of Environmental Studies at Yale-NUS College, but our lived experiences, nodes of personal identity and journey in the environmental movement were vastly different.

On little pieces of recycled paper, we illustrated the themes that we envisioned. The question to guide this visioning was simple: *What is your ideal world?* (Figure 2.1)

Through this self-reflection, it became clear that our personal hopes and visions for the future would become central to the heart of the book – the ecosophy. This initial point of defining our personal identity was integral to ground the experience of collaboration in a way that honoured our individual

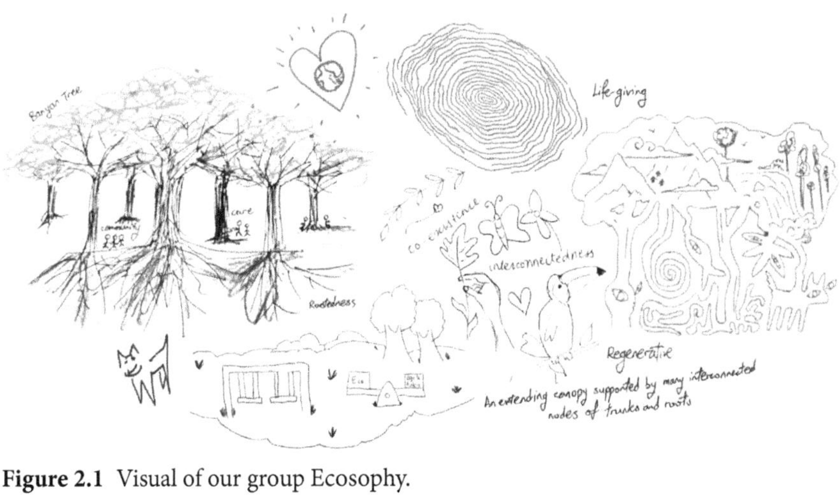

Figure 2.1 Visual of our group Ecosophy.

values but also opened space for them to be changed by being in relationship with each other (Saunders 2008).

Over a span of five sessions, we worked on a shared lexicon of words that portrayed the principles we wanted to see reflected in our ecosophy. On one hand, it became clear that some of our principles overlapped, certain ideas were connected, and that they built on each other. On the other hand, we identified points of tension and ideas that seemed in disagreement with one another and invited discussions of what should be or should not be included in the ecosophy. After all, we should invite all viewpoints, even the possibility of conflicting ones, as part of a pluralist methodology to enable a more holistic, transdisciplinary conversation. These discussions were an opportunity to deepen our collective knowledge as we shared case studies, literature, experiences, worries and hopes to strengthen our ideas.

Ultimately, ecolinguistics as a discipline invokes a love for words and their power to change worlds and climate futures. With this in mind, we played with the vernacular we had grown up with in Singapore and our local linguistic landscape to capture the complex, profound, more-than-life values and meanings that we wanted to convey to ourselves and eventually to our readers. As we brought the discussion down to a simmer, boiling down to what seems to be the essence of our collective ecosophies, we were reminded that this is not the be-all and end-all. It is inevitable that our individual ecosophy will change, but we hoped that this group ecosophy would be a reminder for us to continue discovering, learning and reflecting together. The use of a shared single ecosophy as a framework to conduct ecolinguistical analysis is inspired by Stibbe's methodology, but in no way claims to be a definitive ecosophy, and we encourage you to explore other valid and useful approaches to ecolinguistics too (Stibbe 2014, 10–8). Over one year later, we present this ecosophy to you, the reader:

Ecosophy in One Word: *Majulah…?*

Majulah is the Malay word for 'onwards'. For us, our ecosophy is grounded in the implications of this term:

Community: We aim to expand who we understand to be members of our Singaporean community. Community, to us, includes the human and non-human, marginal groups and transcends state boundaries. Historical and ongoing social and environmental exploitation often stems from the exclusion of certain human and non-human groups. We want to identify spaces that promote the integration and coexistence of diverse actors and plural ways of being and enable the right of all to thrive and to live lives of well-being and dignity. Widening our community also entails that we be inclusive to different ways and forms of knowing and centring those that have been historically underprivileged.

Care: We aim for a future of better living conditions for all in the community by extending compassion, empathy, justice and stewardship to each other, across borders, gender, species, racial, religious and other distinction lines. Caring is recognizing the inequalities in conditions of living, safety and dignity; the exceedance of planetary and societal health boundaries; and the uneven geographies of labour and natural resource extraction. Subsequently, we need to act in response to these embedded inequalities in ways that meet our collective needs but that are also just, regenerative and equitable.

Critical: We aim to progress towards a more just and regenerative future by constantly striving to do better by expanding our worldviews, reimagining the possibilities and reprioritizing our metrics for success in a way that enables the co-creation of more inclusive, resilient and thriving ecofutures. It involves humility in that no one holds monopoly over knowledge and hence should constantly strive to work with diverse actors and with uncertainty. Our ecosophy advocates for a constant and critical reassessment of our positionality and the invisible exclusions and hierarchies that might be embedded in our imagining of a better world. This also includes an awareness of our positionality in writing the book as people living in Singapore, which holds a lot of geospatial privilege and power but also geopolitical complicity in climate conflicts within our region, Southeast Asia. This necessitates our critical lens on Singapore as a text, in order to co-create better visions of the future – its creation being an iterative process that requires constant collaboration and self-reflection.

Majulah…?

Majulah Singapura is the title of Singapore's national anthem, which translates to 'Onward Singapore'. This is a musical and lyrical expression of Singapore's identity as a nation, which is built upon the story of the fishing village turned metropolis. As we elabourated in Chapter 1 on the Singapore Story, this *story-we-live-by* tells a tale of development and growth but at the expense of historical erasure, displacement and environmental degradation. Our ecosophy borrows from the first half of the phrase, to translate to 'Onward…?'. By naming our ecosophy as such, we attempt to *reframe* a dominant narrative that has been the cornerstone of Singapore's identity for the last few decades. We acknowledge this history and confront it with our hope for a better story for us in Singapore, and for the world, in this uncertain future.

The '?' in our ecosophy both reflects the uncertainty the Climate Generation faces but also invites us (both 'us' as writers, and the collective 'us' including you, the readers, as part of the climate-affected world) to question what we define as progress away from the current regime of Singapore's industrial urban ecomodernity. While time moves onwards, we also want to move onwards – away from hegemonic ideologies of growthism and human-centrism and towards a more just and sustainable future. As understood in the strong sustainability perspective that guides this book, the infinite growth of the economic subsystem within a finite and contained ecosystem is impossible. Hence, the '?' at the end of our ecosophy represents the ultimate normative goal of questioning the status quo.

Including the '…' in our ecosophy opens a space of reflection for the plural imaginings of what this onwards could look like. One of the first lessons we learned in our foray into Environmental Humanities is that humans' current hubristic and destructive relationship to nature is not fixed or essentialist to human nature. Instead, it is a product of recent history, dominant culture and the political economy of recent times – it is subject to being questioned, challenged and changed. Other ways of being in relationship with nature have always existed, still exist today and are often subjugated and show glimpses into a reimagined future that we can move onwards towards.

Hence, our ecosophy aims to reimagine what a vision for the future progress of Singapore can be – one that can move towards more sustainable ecological futures. Instead of its historical meaning of progress within Singapore, how can we translate ***Majulah…?*** into a forward movement towards a new world that's ecologically viable for all of us?

Community

'To be one is always to become with many', the feminist philosopher of science Donna Haraway writes (Haraway 2008, 4). This is a key learning from our journey as students and youth in the Climate Generation – that we are not individuals or a species in silos, but rather are interconnected with the larger ecosystem and with the non-human. We draw inspiration from traditional philosophies such as *Ubuntu* which can be translated to 'I am because we are'; *Dao*, which is the process of realizing that human life is inextricably embedded within a larger process of nature (Miller 2003); and the Buddhist law of cause and effect that expounds that practices are connected to their effects across time and persons. These philosophies anchor our systemic approach to understanding the world.

Additionally, we uphold ecocentrism – an ethical belief that individuals, species and biotic communities all have inherent value. We associate this wider ecosystem as equal-footed to human society, and that we share commonality in our diversity. With this realization, we need to 'make kin' – 'to make-with, to become-with, compose-with' with all that are earth-bound (Haraway 2008). Additionally, such relations require an active process of renewal, restoring and persisting relationships among humans and non-humans (Tuck et al. 2012). Anna Tsing writes that this is about 'enabling entanglements', an interactive collaborative process of building relations among humans and non-humans through a multispecies approach (Tsing 2015, 4). By extending our definition of community beyond just human society, we include all in our envisioned futures of well-being for all.

Care

This pillar of our ecosophy adds a normative prescription to what we imagine relationships within the *community* to be like. In this pillar, we want to forefront three intertwined approaches to *Care:* an Ethics of Care approach, a Feminist Decolonial approach and a Degrowth approach.

In seeing a world as being composed of relationships rather than a world populated by isolated individuals, in seeing a world that is held up through interconnectedness that extends across colour, gender, borders and time, we employ an 'Ethics of Care' approach. This term, coined by Feminist psychologist Carol Gilligan, describes a field of ethics in philosophy that centres around care, not just as a virtue, but as a moral imperative (Gilligan 1982). We use Philosophers Tronto and Fisher's definition of *Care* as: 'A species of activity that includes everything we do to maintain, contain, and repair our world so that we can live in it as well as possible. That world includes our bodies, ourselves and our environment' (Fisher and Tronto 1990, 40). We navigate the world, not as an isolated individual, but held up by an intricate web of relationships – family, friends, colleagues, community – indeed, as held up by every human and non-human that maintains the socio-environmental fabric upon which we all live. Without the care that others have endowed upon us and the ecologies of alive places that support our living being, you the reader and we the authors would cease to exist. Hence, in recognizing our interdependence, the act of caring is a moral imperative decision that extends to humans and non-humans alike, to support our collective dignified thriving in the world. Gilligan suggests that ethics of care must be grounded in voice and relationships, in listening and hearing with respect (Gilligan 1982). Tronto further argues that institutional care must consciously recognize the purpose of care, attend to the politics of care and to care in particularistic and pluralistic ways (Tronto 2010). Fundamentally, in an interconnected world, an ethic of care is what we need to connect to one another and to generate possibilities for relating, living and acting collectively – something we desperately need in our present socio-ecological crisis (Puig de la Bellacasa 2010).

In the current world order (including in Singapore), we maintain that there is a need for an explicit anti-colonial, non-hierarchical approach to relationships. A Feminist Decolonial approach aims to create horizontal coalitions of climate solidarity, across borders and hierarchies, to empower front-line communities to seek their own solutions to postcolonial trajectories for climate futures. It recognizes that Most Affected Peoples and Areas/Global South communities are owed climate repatriations and investments from Least Affected People and Areas/Global North communities that help to right (as much as possible) past and ongoing injustices. It also recognizes that degrowth must not threaten livelihoods in developing areas by the withdrawal of Global North/Least Affected Peoples and Areas' finance and jobs.[14] It also foregrounds gender justice and empowers female and gendernoncomforming community members as they take leadership in navigating the 'contradiction[s that] underscore … the complexity and opacity of Singapore's [and the region's] histories, marked by the dichotomies of tradition versus ecology, nature versus culture, and life versus death' (Yoon 2024, 148).

A degrowth approach is grounded in the critique of the growthism and metabolic material throughput of economies that have resulted in ecological overshoot. It is about upturning the modern fixation on economic growth and to reorientate the function of economies to support meeting core basic needs of all citizens, equally. This entails that (1) most global economies, especially those of Least Affected Peoples and Areas/the Global North, need to materially and measurably shrink so that metabolic flows exist within the regenerative carrying capacity of earth's ecosystems; (2) that the success of economics should be measured by well-being metrics that are not based upon gross domestic product (GDP) but instead on quality of life metrics related to happiness, environmental health, education, peace, the provision of basic public goods and services and justice; (3) and that governments have a part to play with policy interventions and planning strategies to make this happen in ways that centre justice, equity and equality while also protecting the environment. This must include environmental protection, accessible green spaces and parks, education, health care, public transit and for some degrowth proponents, a universal basic income and universal housing. We note that currently there is limited research on a combined Feminist Decolonial Degrowth approach in the context of Singapore, and our hope is that future scholars build on what we offer in this book as a starting point.

Critical

While we radically imagine more hopeful climactic futures, we are also at risk of replicating or reifying unjust logics. Hence, a key tenet of our ecosophy is to be critical, even of our radical imaginings of the future. The etymology of 'radical' traces the word back to 'root', hence, with a critical lens, we are constantly tracing back to the roots from which our hopes stem from, and we evaluate whether they come from a place of care and community. Feminist ecological economist Dengler and social scientist Seebacher highlight the importance of interrogating even the most radical ideas according to feminist and decolonial ideals that are anti-hierarchical and do not place subjects into categories based on extrinsically applied and policed value differentials (Dengler and Seebacher 2019). They make a convincing argument that true change cannot occur without resisting heterodox narratives of injustice at every level of the process. Hence, we find it key to a better climactic future to centre discourses that engage us as 'active, society-shaping citizens, rather than consumers, and puts guiding principles such as autonomy and participatory, collective bottom-up decision-making processes at its very core' (Dengler and Seebacher 2019, 248).

A key part of this critical approach is a pluralistic lens to alternative stories and a preference for discourses that allow plural stories to be acknowledged, especially stories from and told by those historically disadvantaged. This is to give voice to the silent actors, such as marginalized groups, non-humans and future generations, into the stories that also impact them. Additionally, we take an ideological stance against violent mechanisms of current, modern logics of extraction and exploitation for urban development, that exterminate these alternate ways of knowing, being and living, and instead propose a climate future of a 'world with many worlds', a Zapatista saying that refers to a world that does not homogenize but allows for the specificities of different communities – both human and non-human – in context to live with dignity (Kothari et al. 2019).

Methods of Ecolinguistical Analysis

Throughout the course of the authorship of this book, many of the authors took their own academic liberties to research, present and analyse their adopted case studies based on both the group ecosophy and informed by their personal interpretations of it. We modified Stibbe's recommendation of ecolinguistical analysis (analysing linguistics first, then evaluating the found stories against an ecosophy) to fit our need for a spatial analysis of a city. We instead started by coming together to identify case studies we felt were most representative of Singapore's diverse environmental purviews, making sure to include both nation-wide common spaces, nature spaces, corporate spaces and residential spaces. All the case studies we felt were representative of a specific aspect of the varied environmental stories of Singapore. The unique spatial element posed a challenge to these authors, as we learnt to look at the spaces we lived in as 'text' and uncover the *stories-we-live-by* that were fruits of this 'text'.

We recognize that some of these 'texts' (stories-as-places) may be unfamiliar to readers, especially to those readers who have never been to the country and may be expecting an analysis of Singapore's tourism hot spots, but as insiders to Singapore, we picked 'texts' that allow for some of the deeper tensions inherent to Singapore to be uncovered and shared. There are many 'gardens' in Singapore, and as Singaporeans and residents of Singapore, we have picked case studies that invite you into some of Singapore's more hidden 'gardens'.

Once decided on, each of the authors took lead(s) on case studies, and then began to accumulate the materials used for analysis. Each case study analysis roots itself in a specific locality (Pulau Bukom, Botanic Gardens) or spatial type (park connectors, agrotechnology hub) (see Figure 2.2 for a map of Singapore and case study locations). Actively undertaking an analysis of both how the space espouses a story or ideology and how residents interact with the space (via group activities, governmental policy and non-human presence) meant that authors collected both field data and other supplementary materials surrounding each space. These supplementary materials ranged from online sources such as government websites to visiting the spaces and photographing aspects of it for analysis. Certain case studies such as Bukit Brown Cemetery and Chek Jawa (Chapter 5) include in their analysis interviews with Singaporean residents.

Figure 2.2 Map of Singapore and case study locations.

Throughout these various collection exercises, the goal of analysis for a deeper understanding of the overall ecolinguistics of Singapore remained.

We invite everyone who interacts with our ecosophy and who shares some part of it in imagining the future progress of Singapore to be similarly driven by visions of a just and sustainable future. In the following chapters, we will be analysing the ideologies about human–nature interactions as these are embedded in the different spaces, discourses and 'texts' that constitute Singapore (as physical space and web of socio-ecological relations) and thus form the stories by which Singaporeans live in this century of globalization and climate change. In analysing these discourses according to our ecosophy, they will be categorized as *destructive*, *ambivalent* and *beneficial* (which follow as Chapters 3, 4, and 5, respectively).

Notes

1 Or those that are considered interdisciplinary, like environmental studies, which all of us as authors study and within which we undertake our respective research.

2 We recognize, however, that a stated ethic is an ideal and that a gap between an expressed ethic or value and actual lived behavior often exists, especially when it comes to environmental and sustainability issues within globalization (Tuan 1970; LeVasseur 2017).

3 Here, we note that the summary sections of environmental ethics of this chapter are written by Todd, while the rest of the chapter is written by its other co-authors. This is also because Chapters 3–5, where the three types of discourses identified by Stibbe are applied to places in Singapore, are done by these same co-authors who generated the ecosophy. Todd, as lead editor and co-author of this chapter, was not involved in generating the ecosophy found later in this chapter or in applying it via ecolinguistical discourse analyses in the case studies of those chapters.

4 We readily recognize that all cultures have what can be considered environmental ethics. We also recognize that land-based and place-based peoples, especially indigenous peoples, typically centre and celebrate *beneficial* stories about their home places that have historically inspired sustainable lifeways, especially in areas where metabolic relations with ecosystems are actively managed for resilience and regeneration (Whyte 2018). Past and ongoing violent processes of colonization and enclosures and the forced relocation and/or education of land-based and indigenous peoples have negatively impacted the respective ecosophical stories of such peoples the world over, including in Singapore.

5 For an ecolinguistical analysis of the Buddhist *jataka* tales, and how the tales can inform an ecosophy, see Bhattacherjee and Sinha (2021). For similar projects, see also Roy (2015) and Kumaravelu and Christopher (2023).

6 This is especially true in Singapore where dengue fever is a very real public health concern, as is heatstroke.

7 For an academic overview of his work, see Best (1998).

8 For an engaging example of this tradition in action in relation to animal ethics, see Donovan and Adams (2007).

9 We recognize the importance of the many years of back-and-forths within all of these approaches to environmental ethics. Given time and space constraints, providing such a thorough history is outside the scope of this chapter.

10 For a rebuttal of Guha's critiques, see Johns (1990).

11 Robin Attfield provides a helpful summary of Sylvan's critiques and of responses to these critiques as offered by Warwick Fox (1993).

12 For a sustained critique of foundational premises and normative assumptions of deep ecology, see Luke (1997), especially pages ten through twenty, and on the political implications of, see pages twenty to twenty-seven. The depth of this critique deserves its own attention from readers interested in possible limits to deep ecology, yet summarizing and articulating this critique is outside the space constraints of this section of the chapter. An important rebuttal to some of the numerous early, 1980s critiques of deep ecology is offered by Kirkpatrick Sale (1988). Another helpful summary of various criticisms of deep ecology is offered by David Keller (2009), especially pages 208–9.

13 The 'we' used in this context refers to the group of co-authors as a whole – Anjali, Beverley, Hannah, Kristina, Lecarl and Yock.

14 This recognizes the import of the economic leg in weak and strong sustainability perspectives.

References

Attfield, Robin. 1993. 'Sylvan, Fox and Deep Ecology: A View from the Continental Shelf'. *Environmental Values* 2: 21–32.

Ban, Cornel, and Mark Blyth. 2013. 'The BRICs and the Washington Consensus: An introduction'. *Review of International Political Economy* 20 (2): 241–55.

Best, Steven. 1998. 'Murray Bookchin's Theory of Social Ecology: An Appraisal of "The Ecology of Freedom"'. *Organization and Environment* 11 (3): 334–53.

Bhattacherjee, Monalisa, and Sweta Sinha. 2021. 'Ecosophy through Jataka Tales'. *Language and Ecology*. https://www.ecoling.net/_files/ugd/ae088a_fa7f36a07ede46c5ba78e4d9089eb75e.pdf.

Bullard, Robert. 1993. 'Race and Environmental Justice in the United States'. *Yale Journal of International Law* 18: 319–35.

Callicott, J. Baird. 2017. 'The Historical Roots of Environmental Philosophy'. In *Religion and Ecological Crisis: The 'Lynn White Thesis' at Fifty*, 33–46. New York: Routledge.

Cheney, Jim. 1987. 'Eco-Feminism and Deep Ecology'. *Environmental Ethics* 9 (2): 115–45.

Chua, Beng Huat. 2017. *Liberalism Disavowed: Communitarianism and State Capitalism in Singapore*. Singapore: National University of Singapore Press.

Costanza, Robert, Jon D. Erickson, Joshua Farley, and Ida Kubiszewski. 2020. 'Introduction: What Is Ecological Economics and Why Do We Need It Now More Than Ever'. In *Sustainable Wellbeing Futures*, 1–15. Northampton, Massachusetts: Edward Elgar Publishing.

Dengler, Corinna, and Lisa Marie Seebacher. 2019. 'What About the Global South? Towards a Feminist Decolonial Degrowth Approach'. *Ecological Economics* 157: 246–52.

Devall, Bill. 1996. 'Response to Bron Taylor's Criticisms of My Review of Ecological Resistance Movements'. *Trumpeter* 13 (3). Retrieved from https://trumpeter.athabascau.ca/index.php/trumpet/article/view/258.

Donovan, Josephine, and Carol Adams, eds. 2007. *The Feminist Care Tradition in Animal Ethics: A Reader*. New York: Columbia University Press.

Fair, Hannah, and Matthew McMullen. 2023. 'Towards a Theory of Nonhuman Species-Being'. *Environmental Humanities* 15 (2): 195–214.

Fisher, Bernice, and Joan C. Tronto. 1990. 'Toward a Feminist Theory of Care'. In *Circles of Care: Work and Identity in Women's Lives*. State University of New York Press.

Gilligan, Carol. 1982. *In a Different Voice: Psychological Theory and Women's Development*. Cambridge, MA: Harvard University Press.
Glasser, Harold. 2011. 'Naess's Deep Ecology: Implications for the Human Prospect and Challenges for the Future'. *Inquiry* 54 (1): 52–77.
Guha, Ramachandra. 1989. 'Radical American Environmentalism and Wilderness Preservation: A Third World Critique'. *Environmental Ethics* 11 (1): 71–83.
Haraway, Donna J. 2008. *When Species Meet*. University of Minnesota Press.
Harris, Melanie. 2016. 'Ecowomanism: An Introduction'. *Worldviews* 20 (1): 5–14.
hooks, bell. 2014 *Teaching to Transgress*. New York: Routledge.
Hutlman, Martin, and Paul Pulé. 2018. *Ecological Masculinities: Theoretical Foundations and Practical Guidance*. New York: Routledge.
Johns, David. 1990. 'The Relevance of Deep Ecology to the Third World'. *Environmental Ethics* 12 (3): 233–52.
Katz, Eric. 2000. 'Against the Inevitability of Anthropocentrism'. In *Beneath the Surface: Essays in the Philosophy of Deep Ecology*, 17–41. Cambridge: MIT Press.
Keller, David. 2009. 'Deep Ecology'. In *Encyclopedia of Environmental Ethics and Philosophy, Volume 1*. 206–11. Detroit: Gale Cengage.
Kothari, A., Salleh, A., Escobar, A., Demaria, F., and Acosta, A., eds. 2019. *Pluriverse: A Post-Development Dictionary*. New Delhi: Tulika Books.
Kumaravelu, Shanmugapriya, and G. Christopher. 2023. 'Eco-philosophy of Indian Classical Fables'. *Ecocycles: Scientific Journal of the European Ecocycles Society* 9 (1): 100–6.
Leopold, Aldo. 1989 (1949). *A Sand County Almanac: And Sketches Here and There*, special edition. New York: Oxford University Press.
LeVasseur, Todd. 2011. 'Bioregion'. In *Green Cities*. Thousand Oaks, CA: Sage.
LeVasseur, Todd. 2014. 'Environmental Philosophy in a Post-Ice Cap North Polar World'. *Environmental Ethics* 36 (3): 303–18.
LeVasseur, Todd. 2017. *Religious Agrarianism and the Return of Place: From Values to Practice in Sustainable Agriculture*. Albany: SUNY Press.
Luke, Timothy. 1997. *Ecocritique: Contesting the Politics of Nature, Economy, and Culture*. Minneapolis: University of Minnesota Press.
Miller, J. 2003. 'Daoism and Nature'. In *Nature Across Cultures. Science Across Cultures: The History of Non-Western Science*, vol 4. Springer, Dordrecht. https://doi.org/10.1007/978-94-017-0149-5_20.
Moore, Jason. 2017. 'The Capitalocene, Part I: On the Nature and Origins of Our Ecological Crisis'. *The Journal of Peasant Studies* 44 (3): 594–630.
Morton, Timothy. 2012. *The Ecological Thought*. Cambridge, MA: Harvard University Press.
Morton, Timothy. 2013. *Hyperobjects: Philosophy and Ecology after the End of the World*. University of Minnesota Press.
Naess, Arne. 1973. 'The Shallow and the Deep, Long-range Ecology Movement. A summary'. *Inquiry* 16 (1–4): 95–100.
Naess, Arne. 1977. 'Spinoza and Ecology'. *Philosophia* 7 (1): 45–54.

Naess, Arne. 1986a. *There Is No Point of No Return*. Penguin, Green Ideas series, #18.
Naess, Arne. 1986b. 'The Deep Ecological Movement: Some Philosophical Aspects'. *Philosophical Inquiry* 8 (1–2): 10–31.
Ortner, Sherry. 2005. 'Subjectivity and Cultural Critique'. *Anthropological Theory* 5 (1): 31–52.
Plumwood, Val. 1986. 'Ecofeminism: An Overview and Discussion of Positions and Arguments'. *Australasian Journal of Philosophy* 64: 120–38.
Plumwood, Val. 1991. 'Nature, Self and Gender: Feminism, Environmental Philosophy and the Critique of Rationalism'. *Hypatia* 6 (1): 3–27.
Plumwood, Val. 1999. 'Comment: Self-Realization or Man Apart? The Reed-Nasess Debate'. *Philosophical Dialogues: Arne Naess and the Progress of Ecophilosophy*: 206–212.
Puig de la Bellacasa, M. 2010. 'Matters of care in technoscience: Assembling neglected things'. *Social Studies of Science* 41 (1): 85–106. https://doi.org/10.1177/0306312710380301.
Rolston, III, Holmes. 1981. 'Values in Nature'. *Environmental Ethics* 3: 113–28.
Rothenberg, David. 1987. 'A Platform of Deep Ecology'. *The Environmentalist* 7 (3): 185–90.
Roy, Varun Kumar. 2015. 'In Search of the Ecosophy: Environmental in Buddhist Philosophy'. *Karatoya: North Bengal University Journal of History* 8: 26–33.
Sale, Kirkpatrick. 1988. 'Deep Ecology and Its Critics'. *The Nation* May 14: 670–75.
Sandilands, Catriona. 2016. 'Queer Ecology'. In *Keywords for Environmental Studies*. New York: New York University Press. https://keywords.nyupress.org/environmental-studies/essay/queer-ecology/.
Saunders, Clare. 2008. 'Double-Edged Swords? Collective Identity and Solidarity in the Environment Movement1'. *The British Journal of Sociology* 59 (2): 227–53. https://doi.org/10.1111/J.1468-4446.2008.00191.X.
Sessions, George. 1987. 'The Deep Ecology Movement: A Review'. *Environmental Review* 11 (2): 105–25.
Sessions, George, and Arne Naess. 1986. 'The Basic Principles of Deep Ecology'. *The Trumpeter* 3 (4): 14.
Seymour, Nicole. 2018. *Bad Environmentalism: Irony and Irreverence in the Ecological Age*. Minneapolis: University of Minnesota Press.
Sheppard, Eric and Helga Leitner. 2010. '*Quo vadis* Neoliberalism? The Remaking of Global Capitalist Governance After the Washington Consensus'. *Geoforum* 41 (2): 185–94.
Stalhammar, Sanna, and Henrik THoren. 2019. 'Three Perspectives on Relational Values of Nature'. *Sustainability Science* 14: 1201–12.
Stibbe, Arran. 2014. 'An Ecolinguistic Approach to Critical Discourse Studies'. *Critical Discourse Studies* 11 (1): 117–28.
Sylvan, Richard. 1985. 'A Critique of Deep Ecology'. *Radical Philosophy* 40. https://www.radicalphilosophy.com/article/a-critique-of-deep-ecology.

Szeman, Imre. 2013. 'How to Know about Oil: Energy Epistemologies and Political Futures'. *Project Muse* 47 (3): 145–68.

Taylor, Bron. 1998. 'Religion, Violence and Radical Environmentalism: From Earth First! To the Unabomber to the Earth Liberation Front'. *Terrorism and Political Violence* 10 (4): 1–42.

Taylor, Bron, and Todd LeVasseur. 2008. 'Ecotage and Ecoterrorism'. In *Encyclopedia of Environmental Ethics and Philosophy*, 286–91. Detroit, MI: Macmillan Reference.

Tronto, Joan C. 2010. 'Creating Caring Institutions: Politics, Plurality, and Purpose'. *Ethics and Social Welfare* 4 (2): 158–71. https://doi.org/10.1080/17496535.2010.484259.

Tsing, Anna Lowenhaupt. 2015. *The Mushroom at the End of the World: On the Possibility of Life in Capitalist Ruins*. Princeton, NJ: Princeton University Press.

Tuan, Yi-Fu. 1970. 'Our Treatment of the Environment in Ideal and Actuality: A Geographer Observes Man's Effect on Nature in China and in the Pagan and Christian West'. *American Scientist* 58 (3): 244, 247–9.

Tuck, Eve, and K. Wayne Yang. 2012. 'Decolonization Is Not a Metaphor'. *Decolonization: Indigeneity, Education and Society* 1 (1): 1–40.

Valbrun, Daphne. 2023. 'Environmental Ethics'. In *Handbook of the Anthropocene*, 1029–33. Springer Nature.

Wang, Xinxin, and Kevin Lo. 2021. 'Just Transition: A Conceptual View'. *Energy Research and Social Science* 82: 102291.

Warren, Karen. 1990. 'The Power and Promise of Ecological Feminism'. *Environmental Ethics* 12 (2): 125–46.

Washington, Haydn, et al. 2017. 'Why Ecocentrism Is the Key Pathway to Sustainability'. *Ecological Citizen* 1 (1): 35–41.

Whyte, Kyle Powys. 'On the Role of Traditional Ecological Knowledge as a Collabourative Concept: A Philosophical Study'. *Ecological Processes* 2.1 (2013): 7.

Whyte, Kyle. 2018. 'Settler Colonialism, Ecology, and Environmental Injustice'. *Environment and Society: Advances in Research* 9: 125–44.

Yoon, Sun Woo. 2024. 'The Life and Death'. In *Lala-Land: Singapore's Seafood Heritage*, 142–51. Singapore: Epigram.

3

Destructive Discourses of Singapore

Beverley Choo Jia Ying and Lecarl Lim

Introduction

Lecarl Lim

The aim of this chapter is to raise awareness of the *destructive* discourses embedded in Singapore and the ideologies they contain. These are stories that are revealed through linguistic and graphic analyses and judged according to the ecosophy of this book. In our ecosophy, we challenge stories that ideologically and discursively perpetuate the alienation of humans from each other and from nature, and that promote economic growth as a fundamental goal of society over socio-environmental well-being. It is important to reiterate that the ecosophy in this book is not definitive, nor are the critiques presented in this chapter the only way to approach the following case studies. Ultimately, we invite you to approach ecologically *destructive* stories critically, to reimagine new stories in their place, and through our ecosophy of *Care, Community* and *Critical*, to imagine just, equitable and eco-sensible ways of being wherever you reside.

Growthist discourse is arguably amongst the most impactful of the various *destructive* discourses present in the two case studies in this chapter. Growthism is the ideology that economic growth is the fundamental goal of society. In Singapore, it has been used to justify the regulation of state and society in the continuance of economic growth in perpetuity. This often entails engaging mechanisms of governance to support economic development and industrialization for rapid economic growth (Tamisari 2017). Singapore's ideology of growthism also attracts regional and international finance, as well as 'unskilled' and highly skilled labour. At the core of the pursuit of the growth ideology and narrative is the assumption that a continual increase of gross domestic product (GDP) will serve and benefit more people by improving living standards, increasing consumption of consumer goods, and generating higher wages, often through increasing production of goods, services and revenue. For example, in Singapore's Committee of Supply Debate 2024 speech titled 'Grow our Economy', the Minister for Trade and Industry Gan Kim Yong (2024) said:

> We can seize new opportunities for growth, improve the lives of all Singaporeans, and build our shared economic future together.

However, as shared in the introduction and from a strong sustainability perspective, this restless pursuit of economic growth is dangerous because infinite growth on a finite planet will only accelerate environmental degradation. Ecomodernist Sustainable/Green Growth proponents assert that economic growth can be pursued while meeting environmental sustainability goals. This decoupling of environmental impact from economic activity is argued to be possible with technological change and substitution. However, even with the most hopeful technological innovations, permanent decoupling is impossible because gains in resource efficiency are ultimately governed by physical limits (Ward et al. 2016). In other words, we can innovate the car to maximize material and fuel efficiency, but the glass, metal and fuel needs cannot be fully eliminated, nor the need for urban planning around roads. Even if there are gains in resource efficiency, as Foster, Clark and York point out nicely, 'an economic system devoted to profits, accumulation, and economic expansion without end will tend to use any efficiency gains or cost reductions to expand the overall scale of production' (Foster, Clark and York 2010). Hence, continued economic growth will only increase the ecological footprint of an economy (Hickel and Kallis 2020). After all, as ecological economist Herman Daly succinctly puts it, 'the economy is an open subsystem of a finite and nongrowing ecosystem' (Daly 1991). Yet the successful and timely decoupling of resource use from growth in GDP seems highly implausible, especially to keep us below 1.5°C and where this insight supports our normative position of Feminist Degrowth. It would therefore also be misleading, maladaptive and ecologically destructive to develop a growth-oriented policy that is predicated on the expectation that decoupling is possible (Hickel and Kallis 2020).

The costs of the ideology of growthism are not just ecological but social too. Firstly, the benefits of economic growth are not equally shared and often lead to higher wealth inequality because of the disproportionate ceaseless accumulation by capital-owners. According to the UBS Wealth Report 2024, Singapore has only become more unequal, with the Gini coefficient rising 22.9 per cent since 2008 to seventy points in 2023 (Tan 2024). Secondly, Singapore's model of progress, to achieve economic growth, has been to aim for productivity. However, we can only be so productive before we feel physiological limits to the work and stress we can handle. Singapore has been surveyed to be the most fatigued country in the world, outdoing China and Japan (Lay 2021). When we are overworked, it also becomes more difficult to care for family and friends, to pursue passions, to stay healthy and to be part of (more-than-human) communities. Therefore, the ideology of growthism, and its close cousin, productivism, runs counter to the ecosophy of this book because it decentres care for our communities and environment to favour the pursuit of growth.

Hence, for Singapore, we will need to confront the fact that economic growth will increase pressure on our already stretched land, labour and carbon resources. Therefore, to resist growthist discourses would be to question growth for growth's sake and set us to look beyond growth for alternatives.

Another notable *destructive* discourse is anthropocentrism, which is the ideology of human-centredness and human exceptionalism. From early capitalism through to the Industrial Revolution, colonization of nature and frontier movements have been justified by the notion that humans are conquerors of the land, and hence can do whatever they want to it (Moore 2017). Anthropocentrism finds its roots in the stories of an assumed and policed human–nature binary whereby human and non-humans are communicated as separate (Milstein et al. 2023). This separation is often justified by arguments that emphasize human intellectual superiority or divine selection, thus positioning humans and our needs as more valuable than non-human needs. Hence, what is valuable to humans is prioritized and preserved, and what cannot be valued by humans often risks systematic erasure.

Erasure often brings 'certain elements together into a configuration while leaving out a whole universe of other elements' (Stibbe 2015, 146). This means that entire aspects of the non-human world can be excluded from our collective awareness. What makes this phenomenon truly *destructive* is that one cannot care about the things that are left out of a discourse because this absence and loss may go unnoticed and unchallenged. For example, natural habitats are difficult to value in human-centric terms and are thus often erased in decisions to pursue logging and mining. As a result, the devastation of natural habitats can be framed as rational and even just in pursuit of human advancement. In short, this process of *erasure* enables these stories that render the unconscionable exploitation and destruction of non-humans and ecosystems not only possible, but morally acceptable. In Singapore, the *erasure* is of nature, as evidenced in how nature has systematically been erased from Singapore's history textbooks that were published between 1984 and 2015. According to the environmental communications, literature and humanities scholar Matthew Schneider-Mayerson and the independent scholar Xiaoyun Neo, 'We find that Singapore's nationally mandated history textbooks represent the nonhuman environment via utilitarian, negativistic, and dominionistic values, and that the very existence of the nonhuman environment is largely relegated to narratives about Singapore's distant past, disappearing almost entirely from narratives about modern, post-colonial nation-state' (2021, 61).

By representing the non-human environment in the above manner, anthropocentrism also finds its way into well-meaning Western-origin conservation efforts that view 'nature' as a subject separate from human spaces, and to be protected (Cronon 1991). Well-demarcated and enforced nature parks become zones to be devoid of human activity (and often created by removing original indigenous dwellers), while human spaces are permitted full exploitation of unprotected ecologies. Anthropocentrism is a paradigm that currently dominates societal norms with serious implications to the worsening ecological crisis, as argued by Aldo Leopold back in 1949 and Arne Naess in the 1970s and 1980s. Hence, anthropocentric discourse is destructive according to our ecosophy.

The following two ecolinguistical analyses of 'texts' in Singapore demonstrate how growthism and anthropocentrism are central to the *destructive* stories being told in these Singaporean spaces. In the first case study, the construction and continued support for the fossil fuel and petrochemical sector in Pulau Bukom, Singapore, demonstrates the support for fossil fuel-driven economic growth, and anthropocentric sidelining of environmental harms during its construction. This has been justified through two ecolinguistics discourses:

1. A *conviction* discourse that increases the facticity of fossil fuel and petrochemical industries as positive contributors to socio-environmental outcomes, rather than being viewed as sources of environmental degradation.
2. The *erasure* of the effects of the fossil fuel and petrochemical sectors, leading to a lack of awareness about the broader consequences of these industries.

In the second case study, Singapore's agri-food story clearly exemplifies growthism, characterized by a commitment to industrial economic growth at the expense of food self-sufficiency. The government's ambitious agrotechnological initiatives within the agro-food sector reflect growthist and anthropocentric ideologies, which can be unpacked through three key ecolinguistic discourses:

1. *Euphemization* strategies around agrotechnological ambitions that obscure the exploitative practices involved and that perpetuate a distant human–animal relationship.
2. *Appraisal* patterns that privilege modernity over tradition, leading to a disregard for traditional ecological knowledge and diminishing its relevance in contemporary discourse.
3. *A conviction* discourse used to justify the control of nature as both necessary and sustainable, shaping public perception to align with these growth-centric narratives.

We now analyse two specific case studies that illustrate how these *destructive* discourses and their underlying ideologies are present in two 'texts' of Singapore.

Pulau Bukom: Singapore's Petroculture Past and Present

Beverley Choo Jia Ying

But Singapore accounts for only 0.1% of global emissions!

This is a statistic repeated by Singapore's leadership at almost every climate and environmental event. It is found on various government websites and was cited in speeches by Minister for Sustainability and the Environment Grace Fu at the National Energy Efficiency Conference in 2021 (Fu 2021), Minister for Finance Lawrence Wong at Singapore International Energy Week in 2022 (Wong 2022), Coordinating Minister for Economic Policies Heng Swee Keat at the launch of the SGX Sustainability Reporting Review in 2023 (Heng 2023), and perhaps most recently as this book went to press, by Singapore's Ambassador for Climate Action Ravi Menon at the launch of the COP29 Singapore Pavilion in Azerbaijan on 12 November 2024 (Menon 2024). Oftentimes, the '0.1%' figure is mentioned to highlight just how much Singapore is doing to combat climate change – that we are doing more than our fair share to mitigate the worst effects of the climate crisis.

While this statistic may be true, Singapore's per capita emissions tell a different story. In 2022, Singapore ranked twentieth out of 156 countries or territories in terms of carbon dioxide (CO_2) emissions per capita (International Energy

Agency n.d.), and second highest in the world for energy consumption per capita (CIA n.d.). If we take a closer look at where Singapore's emissions come from, we find that around one-third of Singapore's total CO_2 emissions are direct emissions from its refining and petrochemicals sector (Singapore Parliament 2024). Singapore has long been a major hub for oil refining and petrochemical production, with the energy and chemicals industry accounting for a quarter of Singapore's total manufacturing output in 2022 (EDB n.d.). Refined oil gets exported to other nations for their energy needs, while petrochemicals (substances obtained by the refining and processing of petroleum or other fossil fuels) are used to create various products, from food preservatives and cosmetics to fertilizers, and perhaps most notoriously and ubiquitously, plastics.

The city's dependence on oil and petrochemicals is a narrative that is too often *erased* – until the black sheen of an oil spill forces residents to confront the *salience* of oil for Singapore. On 14 June 2024, Singapore experienced its worst oil spill to date when more than 400 tonnes of oil were spilled across the southern coast of the island, reaching Sungai Rengit, Malaysia, east of Singapore's international border (*The Straits Times* 2024). The oil spill brought to the fore how oil has pervaded life in Singapore and how Singapore is a key global node of petroculture (Wilson, Carlson and Szeman 2017). Singapore, the green, garden city, the glittering technocratic gem with its well-managed green and blue spaces, masks such stains of oil through strong, state-curated narratives around greenery, environmentalism and the greenwashing of its large petrochemical refineries and its reliance on dense forms of energy needed to keep it shining. To understand how this came to be, we need to rewind to the British colonial era, where Singapore's relationship with fossil fuels first began on a small, offshore island located five and a half kilometres from the mainland called Pulau Bukom (Cornelius-Takahama and Loo 2016).

History of Pulau Bukom

Pulau Bukom's history with the fossil fuel and petrochemical industry dates back to the 1880s when the island was used to store packed kerosene (The Shell Company of Singapore Limited n.d.). Then, in 1891, the government rejected a proposal by Syme and Co., representatives of London-based company Marcus Samuel and Co., to build petroleum bulk storage tanks on mainland Singapore. Instead, the tanks were permitted to be built on Pulau Bukom the following year as what would be the first-ever petroleum bulk storage installation in the East (*Singapore Trade and Industry* 1966). In 1897, Marcus Samuel and Co. was renamed as Shell Transport and Trading, and in 1907, they merged with the Royal Dutch Petroleum Company to form Royal Dutch Shell – more widely known as just 'Shell' – one of the world's major energy companies today (Shell n.d.).

Fast forward to December 1959 when Shell announces plans to establish a refinery on Pulau Bukom (*The Straits Times* 1959b). At this point, Shell had already turned Bukom into 'the biggest centre for oil storage, blending, packing and bunkering in Southeast Asia, and one of the largest in the world' (Cornelius-Takahama and Loo 2016). Construction began on 14 June 1960 and was completed in record time. In thirteen months, the refinery was built and officially declared open by then Finance Minister Goh Keng Swee on 26 July 1961 (Cornelius-Takahama and Loo 2016). This behemoth of a project known as Shell Bukom was the world's second largest refinery complex in the 1970s.

Upon its opening, the refinery complex was contained within Pulau Bukom Besar. However, as the fossil fuel and petrochemical industry continued to expand on Pulau Bukom in the years to follow, growing ambitions meant a growing need for more land. In 1970, 360 families living on neighbouring Pulau Bukom Kechil were made to relocate to make way for oil tanks, expanding the Shell Bukom project – which employed 1,000 people at this point (Campbell 1968) – even further. Today, the petrochemical complex on Pulau Bukom includes Singapore's Southern islands of Bukom Besar, Bukom Kechil, Pulau Ular and Pulau Busing, joined together through a gradual process of land reclamation (Wai 1982).

It wasn't just Bukom that was undergoing such development, however – three new refineries were built on other Singapore Southern Islands in the 1970s, including one built by the Singapore Petroleum Company on Pulau Merlimau in 1971; another on Pulau Ayer Merbau in a joint venture between the Singapore Government and Sumitomo Chemical Corporation of Japan in 1974; and one on Pulau Pesak by Mobil in 1977 (Centre for Liveable Cities 2021). Then, in 1982, the government began feasibility studies to amalgamate the islands of Pulau Merlimau, Pulau Mesemut Darat, Pulau Memesut Laut, Pulau Meksol and Pulau Seraya to 'enhance the expansion of petroleum and petrochemical manufacturing and related activities as well as Singapore's entrepot trade' (*The Straits Times* 1982). After more land reclamation, Pulau Merlimau and Pulau Seraya – in addition to Pulau Ayer Merbau, Pulau Sakra, Pulau Pesek Kecil, Pulau Pesek and Pulau Ayer Chawan – were merged to become what Singaporeans know of today as Jurong Island.

As highlighted by Aiden Mock in his chapter in *Eating Chilli Crab in the Anthropocene* (2020), the fossil fuel and petrochemical industry – and Shell's investment in Bukom specifically – played a crucial role in helping newly independent Singapore find its footing and achieve not just growth, but a sense of security. Shell's refinery on Bukom was Singapore's biggest foreign direct investment at the time (Ng 2013, 11), and the company continued to invest significantly in the economy in the years to come. Between 1960 and 1976 alone, Shell invested roughly $65 million per year in Singapore through manufacturing, installation and marketing facilities (Shell Singapore 1976, 22), and the standard of living on Bukom was reported to be better than its neighbouring islands (*The Straits Times* 1959a). Shell had invested in infrastructure beyond its pipelines and storage units on Bukom (Shell Singapore 1976) – they built four community clubs for sociocultural enrichment, an open-air cinema that screened three films weekly, two primary schools attended by 1,000 students, a clinic that gave free medical attention to not only Shell employees and Bukom residents, but residents of neighbouring islands too (Fang 1961), and more.

However, while Shell's investments in Bukom contributed significantly to Singapore's economic growth and in a number of ways to its social well-being, the company's expansion in the city-state did not come without sacrifices.

Figure 3.1 Pulau Bukom as visible from a boat on the way to Pulau Hantu, with its flat terrain, storage tanks and towering structures. (Photo credit: Hannah.)

Destructive Discourse around Bukom

Community and Care: What about the Corals?

While Shell's efforts to create a liveable environment for their workers on Bukom are laudable, it fails to resonate with our ecosophy's demand for Community and Care for the non-human. When Shell needed more land for the Bukom Refinery in 2005, an Environmental Impact Assessment (EIA) was commissioned by government-linked JTC Corporation to determine the potential effects of land reclamation on the ecosystems surrounding Pulau Ular. This was pertinent given Pulau Ular's proximity to Pulau Hantu, a popular local dive site high in marine biodiversity. Also notable is the fact that, before it was reclaimed, Pulau Ular was a submerged coral reef with just two visible islets (*The Straits Times* 1974).

When news of the reclamation project broke to the public, many had concerns. Local marine conservation groups Blue Water Volunteers and Hantu Bloggers wrote to Shell to express their concerns about the impact of the reclamation project on marine biodiversity, and their voices were amplified by local newspaper *The Straits Times* (Basu 2005). With such concern and pressure from the grassroots level, this issue was even brought up in Parliament on 20 July 2005 (Singapore Parliament 2005).

Ultimately, the EIA made a number of recommendations to enable 'minimal long-term impacts of land reclamation activities on the surrounding marine environment' (DHI n.d.). This included relocating over 3,500 coral colonies from Terumbu Bayan reef – an undisturbed coral reef with high coral cover and rich biodiversity located close to Pulau Ular – to other sites in Singapore before reclamation work began (DHI n.d.). This was just 35 per cent of the live hard coral cover of the reef. The reclamation project was eventually completed in February 2007. The marine biodiversity community grieved the loss of Terumbu Bayan's unique ecosystem, and since no long-term follow-up studies were conducted, it is unclear how many of the relocated corals even survived the stressful move. There is also no publicly available information about Pulau Ular's own reef and the biodiversity that was lost after they were smothered by land for Shell's greater Bukom project. Here, *erasure* in ecolinguistic theory[1] manifests itself not just linguistically – in the form of a story told by both Shell and the government that local biodiversity and natural environments are unimportant in the face of fossil-fuelled development – but also physically as coral reefs, forests and entire hills were quite literally erased from Bukom: this is perhaps the very definition of a *destructive* discourse.

In fact, Singapore has already lost half of our coral reefs to land restoration alone (Chou 2023). As a scuba diver, I've had the privilege of witnessing the beauty and diversity of our marine life, and truly, each coral reef ecosystem is unique. These reefs are not just aesthetically pleasing; they are vital ecosystems that support countless marine species and protect our coastlines (Harris et al. 2018; Ng and Chou 2018). For Singapore's Indigenous communities, the marine life along Singapore's shores are also deeply meaningful to their culture and heritage, as many lived on the coast and relied on various reef species for food prior to the government's decisions to relocate islanders to the mainland and to ban fishing and foraging (read more about this in Hannah's case study on Chek Jawa in Chapter 5). This adds nuance to the anthropocentrism displayed by the Singaporean government in their efforts to expand the fossil fuel and petrochemical industry in the country. Specifically, in the government's relentless pursuit of petro-modernity, they were willing to not only forsake non-human life in the form of destroyed coral reefs and felled trees, but human culture too, as Southern islanders' and Indigenous communities were forced to relocate and lose their intimate connection with the land, sea and its ecosystems as a result.

Singapore's economy is nowhere near as small and fragile as it was when Shell made its first investment in Bukom over sixty years ago. While Shell's contributions to Singapore's economy are undeniable, the sacrifice of our already-threatened and delicate ecosystems for economic gains – let alone to grow the fossil fuel industry that is driving ongoing and worsening anthropogenic climate change – is a trade-off that we should no longer entertain. The ideologies of growthism and anthropocentrism have determined the biophysical 'texts' of Singapore in ways that are *destructive*, and that run counter to our call for degrowth in an era of runaway climate change.

Critical: Imagining Futures without Fossil Fuels

The final pillar to our ecosophy is 'Critical', which urges a 'constant re-assessment of our positionality and the invisible exclusions and hierarchies that might be embedded in our imagining of a better world'.[2] The stories told by Shell and the Singaporean government make it difficult to imagine a world without the fossil fuel industry. By using what Tilsted et al. (2022) call 'strategic narratives', both Shell and the Singaporean government position the petrochemical industry as an imperative and almost permanent fixture of our future – not in spite of the climate crisis, but because of it.

Tilsted et al. (2022) identify three strategic narratives that the largest petrochemical producers adopt to justify their present and future existence in light of the climate crisis. Specifically, these narratives are: (1) Realizers of sustainability; (2) Breakthrough technology pioneers; and (3) Already well underway. Each of these strategic narratives are *framing tools*[3] that the petrochemical industry uses to 'accommodate disruptive pressures and position themselves as transition enablers that are key to successfully achieving decarbonization and circularity' (Tilsted et al. 2022, 2). With this framing, fossil fuel companies tell a story to form a *conviction*[4] in the minds of consumers that their products are essential to human and planetary well-being.

The 'realizers of sustainability' narrative highlights how various petrochemical products are ecologically beneficial, such as plastic packaging production being less carbon-intensive than glass or metal packaging. Such *framing* positions the industry as an ally and enthusiastic participant in the race to carbon emissions reductions, despite the fact that fossil fuels are used as a crucial feedstock for petrochemical products. For the 'breakthrough technology pioneers' narrative, petrochemical producers position themselves as key drivers of low-carbon technologies such as carbon capture (utilization) and storage (CCS/CCUS) and hydrogen (feedstock and power generation). This narrative implies that we must necessarily continue to fund and expand the petrochemical industry, lest we fail to develop and deploy these critical technologies. Finally, the 'already well underway' narrative attempts to present the petrochemical industry as taking and having already taken concrete action to advance climate action, while framing their continued existence as being essential to the ongoing ease and comfort of human lifestyles.

It is important to note that Tilsted et al. point out key differences between fossil fuel extractors and petrochemical producers, arguing that petrochemical producers are able to frame themselves as enablers of an ecomodernist future – a privilege that the fossil fuel industry does not have: 'Ultimately, oil and gas firms face an existential threat, and operate with the fundamental options of either not remaining oil and gas firms or not remaining in business' (Tilsted et al. 2022, 9). However, since Shell constructed the Shell Eastern Petrochemicals Complex (SEPC) in 2010, which spans Pulau Bukom and Jurong Island, Shell Singapore is both a fossil fuel extractor and refiner *and* petrochemical producer and can thus employ – and does indeed employ – the strategic narratives outlined in Tilsted et al.'s paper.

The multi-billion-dollar SEPC was Shell's largest petrochemicals investment (Shell 2010) and was expected to attract $2 billion in fixed asset investments to Singapore from other chemical firms (Wong 2010). Indeed, the SEPC cemented Singapore's role as a Southeast Asian petrochemical 'superpower' given the strategic import of Shell's investment and the operational logistics of its petro-ecologies and material contribution to global petroculture. Now renamed as the Shell Energy and Chemicals Park Singapore, its website reads:

> Situated on Pulau Bukom, it is Shell's only energy and chemicals park in Asia. What was once an oil storage installation and later Singapore's first refinery in 1961, has transformed into an energy and chemicals park that will focus on producing low-carbon energy products like biofuels; incorporate circularity, such as waste plastics for feedstock; as well as provide renewable energy. (Shell Singapore n.d.)

With such *framing*, Shell suggests that fossil fuel activities are a thing of their past, and they have pivoted instead to producing biofuels and renewable energy (a small instance of the 'realizers of sustainability' narrative) and suggest they are committed to sustainable practices like using waste plastics for feedstock (a small instance of the 'already well underway' narrative).

Yet, since taking the helm in January 2023, Shell's new CEO Wael Sawan was clear about his plans to boost the company's profits by increasing natural gas production, steadying oil production and focusing on higher-margin projects. What this has looked like at the time of writing has been Shell pulling out of many wind, solar and low-carbon ventures (Bousso 2024), scaling back hydrogen (Bousso 2023) and watering down Shell's emissions targets globally. In March 2024, Shell completely scrapped a goal to reduce its carbon intensity by 45 per cent by 2035 and weakened its original goal of reducing the net carbon intensity of its energy products by 20 per cent by 2030 to a reduction of 15–20 per cent instead (*CNN* 2024). A study published in 2022 examining the extent of decarbonization and clean energy transition activity of fossil fuel giants BP, Chevron, ExxonMobil and Shell also found that, despite an increase in keywords in annual reports related to climate change and clean energy, the fossil fuel producers had empty pledges and displayed business strategies and investment behaviour that are 'sharply misaligned with tendencies toward increased green discourse' (Li, Trencher and Asuka 2022, 19). Ultimately, because they are liable to their shareholders, leaders of fossil fuel and petrochemical companies have been making decisions that maximize profits over all else, even in light of decades of ever-worsening climate change. Ironically, their ongoing operations fuel the very same climate change that threatens the petro-ecologies of Singapore with sea level rise – Singapore's mean sea levels are projected to rise up to 1.15 metres by the end of the century (National Environment Agency 2024).

Despite this, the Singaporean government fuels and reinforces the strategic narratives put forth by Shell and the local petrochemical industry as a whole and insists on growing the industry as part of its 'green growth' efforts. One of the five pillars of the Singapore Green Plan 2030, which lays out the government's agenda on climate action and sustainable development, is 'Green Economy'. This pillar expresses the government's intention to 'seek green growth to create new jobs, transform our industries and harness sustainability as a competitive advantage' (Ministry of Sustainability and Environment 2024). Singapore's belief in green growth and ecomodernism has long been established, including earlier in this book. Notable here, however, is the government's explicit mention of the petrochemical industry and its importance in realizing their goals: 'The Energy and Chemicals sector is also a key partner in the development of emerging low-carbon technologies, such as carbon capture, utilization, and storage. These technologies can be pivotal in enabling decarbonization at scale, and are crucial in helping us achieve our long-term climate ambitions' (Ministry of Sustainability and Environment 2024). On the surface, this may be read simply as a salutary gesture by the government. However, with an eco-critical lens, we find that this statement only reinforces the *conviction* that the fossil fuel and petrochemical sectors have been perpetuating as a key narrative amongst the Singaporean population – that these industries are enablers of greater socio-environmental outcomes, rather than (or at the very least, this overshadows their roles as) enablers of socio-environmental degradation through the acceleration of the climate crisis.

Of course, not all energy and chemical firms in Singapore engage with fossil fuels. However, given the reference to CCUS, it can be reasonably assumed that they are referring to the fossil fuel and petrochemical giants like Shell. In fact, the Singapore Economic Development Board (EDB) signed a Memorandum of Understanding with ExxonMobil Asia Pacific and Shell Singapore in December 2023 to develop a cross-border CCS project capable of capturing and permanently storing at least 2.5 million tonnes of carbon dioxide (CO_2) a year by 2030 (EDB 2024). In a press release, Shell's Global CCS Vice President Bernhard Koudelka said, 'As the lead CCS developers with ExxonMobil, we are proud to contribute to Singapore's decarbonization plans. With our global track record and deep expertise in developing CCS hubs with over 10 million tonnes of carbon dioxide captured to date, we have the experience needed in bringing together multiple partners and emitters to establish a complex cross border CCS network' (Shell Singapore 2024). Here, Shell reasserts their position as a key collaborator in the fight against climate change by utilizing the 'breakthrough technology pioneers' narrative. But, what good is a project that can store 2.5 million tonnes of CO_2 per year when Shell's direct greenhouse gas emissions from the facilities they operate (Scope 1 emissions) in 2023 alone was 50 million tonnes of CO_2e (Shell Global n.d.)? While they might be investing in decarbonization technologies, being a vertically integrated fossil fuels and petrochemicals company means that Shell is, at the end of the day, a key facilitator of carbon lock-in (Tilsted et al. 2022, 6) that is directly opposed to our ecosophy which envisions a healthy and regenerative future for all beings.

At the time I am writing this, scientists are 'virtually certain' that 2024 will be the first year to be more than 1.5°C hotter than in the preindustrial era (Thompson 2024), the Philippines just experienced an unprecedented extreme weather event of six devastating typhoons in the span of one month (IFRC 2024), and the inaugural planetary health check found that we have breached six (almost seven) of the nine planetary boundaries (Planetary Boundaries Science 2024). The triple planetary crisis of climate change, pollution and biodiversity loss is accelerating at a pace we had not anticipated, and the fossil fuel industry is a key culprit. While CCS and other geoengineering efforts can play a supporting role in climate change mitigation, they absolutely cannot replace the urgent need to reduce greenhouse gas emissions through renewable energy adoption, cutting excessive and unnecessary emissions and transforming our societal systems that are currently built on continued extraction and increased consumption. Without a rapid and comprehensive transformation in this manner, climate change will only further threaten Singapore and its future on a post-1.5°C warmed planet.

Imagining a Post-Shell Bukom

Access to Pulau Bukom is highly restricted. Members of the public are not able to set foot on the island, so many will never interact directly with it. For all intents and purposes, it is *erased* from our lives as Singaporeans, or at least from the humans of Singapore. Yet, there is not a single person on the planet untouched by the products generated at Bukom (and the countless places like it around the planet). The plastic container you use for take-away food at the hawker centre, the fertilizer used to grow the vegetables you had for dinner, the tires of your car and even the laundry detergent *and* polyester shirts in your washing machine, they are all petrochemical products. The microplastics now in human brains, hearts, testicles and even breast milk are all petrochemical products. Those who live in Singapore's southwestern neighbourhoods also have the additional experience of getting used to the night sky never really getting too dark – instead of stars, the night sky is constantly lit up by an ominous, inconsistent glow emanating from the gas flares at Bukom and Jurong Island. Even so, many don't know where this glow comes from.

As the classic saying goes, 'out of sight, out of mind'. Or as ecolinguistics says, this is *erasure*. With Bukom made so inaccessible and petrochemical products becoming so insidiously ubiquitous, fossil fuels have become embedded in our subconscious as a normalized, unquestioned part of our daily lives. They are central to the *stories-we-live-by*, including the *destructive* stories of growthism and anthropocentrism. But, per our ecosophy and the normative goals of ecolinguistics as an interdisciplinary field, this is unacceptable. As the climate crisis intensifies, claiming more lives and irreplaceable ecosystems, we can no longer afford to live in (wilful) ignorance and fall victim to the industry's (and Singaporean government's) distracting strategic narratives. According to the International Energy Agency, petrochemicals are set to account for over a third of the growth in oil demand by 2030 and nearly half of oil demand by 2050 (International Energy Agency 2018). There is no reconciling degrowth, let alone decarbonization and climate action, with supporting the growth of the petrochemicals industry. This is as true for Singapore, the 'green' City in Nature, as any other urban area the world over.

Disruptive narratives and events like the oil spill in June 2024 can visually cue how harmful the city's reliance on the fossil fuel and petrochemical industries can be. However, for many Singaporeans, the *salience* of the harms of these industries disappeared along with the oil spill once it was cleaned up. Petroculture has and continues to shape the social lives, discourses and imaginaries in Singapore. Ecolinguist Arran Stibbe wrote that, 'The way of dealing with destructive discourses is through resistance. Resistance consists of raising awareness that the ideology conveyed by the discourse is just a story, and that the story has harmful effects' (Stibbe 2015, 28). Already, there have been several attempts to shed light on the contradictions and harms perpetuated by Shell and the wider fossil fuel and petrochemical industry in Singapore. One youth-led example is the local student-led divestment movement Students for a Fossil Free Future (S4F). In 2022, S4F published a report that lays bare how the fossil fuel industry has deep ties with Singapore's universities. From scholarships and prizes (e.g. the Shell Medal and Prize awarded to students in the NUS Master of Science (Environmental Management) programme) to sponsoring campus events (e.g. ExxonMobil has been sponsoring the ExxonMobil Campus Concerts series at NUS since 1986), the industry pervades Singapore's higher education landscape in seemingly harmless ways. However, as S4F rightly points out, these actions are strategic: 'By [undertaking these endeavours], they have gained broad social acceptance and approval as partners in developing Singapore's education and economic landscape – In other words, they have purchased a "social license" from our universities to continue their status quo activities' (S4F 2022, 13). By organizing various student engagement events and speaking with the press to raise awareness of their findings and calls to action, S4F built *salience*[5] about the detriments of the fossil fuel industry and the complicity of Singapore's higher education institutions in this by making local universities' invisible ties to the industry visible and mobilizing Singaporean society (specifically university students) to take action.

As we critically imagine a (Singaporean) fossil-free future, we might ask: What about Bukom? Well, there was a Bukom before Shell, and there will be one after. In fact, this might be a reality soon. In the middle of 2024, Shell's press team announced that it sold its energy and chemicals park – which includes Singapore's very first oil refinery that Shell built in 1961 – to CAPGC, a joint venture comprising Indonesia's Chandra Asri Capital and global commodities trader Glencore (Subhani 2024). But, despite Shell's departure, it is evident that Bukom will continue to be a fossil fuels and petrochemicals hub. At the opening of ExxonMobil's expanded Singapore Chemical Plant in 2014, Prime Minister at the time Mr. Lee Hsien Loong made sure to 'assure all the energy and petrochemicals companies here that the Singapore Government stands fully behind them and will continue to help them to succeed' (Lee 2014). A decade has passed since the former Prime Minister's statement, and surely, a reconsideration is in order.

Writing about the 'corporation is a person' metaphor promulgated by the discourse of neoclassical economics, Stibbe (2015, 77) warns us that, 'a danger of representing a corporation as a person who decides things for itself is that it becomes very difficult to locate responsibility for behaviour that damages people's wellbeing and the environment'. We must remember that behind the strategic narratives and ecologically destructive decisions advanced by the fossil fuel and petrochemical industries are *people* – powerful and exorbitantly paid people like Singapore's ministers (Buchholz 2024) and Shell CEO Wael Sawan. We must hold each other – especially our elected government officials – accountable to ensure a liveable future and create new stories-we-live-by that are lucid to the fundamental incongruity of fossil-fuelled industries with said future.

Prior to reclamation, Pulau Bukom hosted a hilly terrain hugged by mangroves and coral reefs (Campbell 1971). Today, its flattened topography is weighed down by massive storage tanks and towering metal structures that process up to 500,000 barrels of oil per day (Shell Singapore 2019). Perhaps it is time for another transformation. Bukom's fossil fuel facilities were pivotal in Singapore's early days, but as our economy has matured and the climate crisis has only grown more urgent, we ought to transition Pulau Bukom away from fossil fuels *and* petrochemicals. We cannot transition away from fossil fuels without also scaling down and eventually phasing out the production and consumption of petrochemicals. It is only once we pursue such a transition – in a manner that is just and prioritizes fossil fuel and petrochemical workers – that we can truly say that Singapore is doing our fair share to mitigate the worst effects of the climate crisis. Until such a time as this, then the 'text' of Bukom island will remain one of Singapore's most *destructive*, telling a story of growthism, anthropocentrism, and one that sees fossil fuels and petrochemicals as central to the well-being of Singapore.

Singapore's Agri-Food Industry: Critiquing Ecomodernism

Lecarl Lim

On 27 October 2024, 106 live quail were discarded in a rubbish bin outside Singapore's only licensed quail farm. The Co-CEO Anbarasi Boopal from The Animal Concerns Research and Education Society (Acres) remarked 'We did not count the lifeless chicks and unhatched eggs because there were so many, possibly in the hundreds' (Shabana Begum 2024). For many in Singapore, this news evoked deep sadness and a sense of injustice over the farm's flagrant disregard for animal life (r/singapore 2024). However, this instance of malpractice is merely a scratch on the surface of a cruel and environmentally dangerous agri-food industry that is only accelerating in voracity.[6]

But despite the well-documented harms of the factory farming industry, how does it continue to justify itself? Stibbe argues that it does so through the discursive strategies offered by the factory farming industry to justify, sustain and expand its practices (Stibbe 2015, 27). He cites Glenn's analysis of how linguistic devices are used to construct animals as commodities, that factory farming benefits animals, and that animals are objects (Glenn 2004). Additionally, Trampe highlights that within industrial discourses, 'living beings are in accordance with the economic-technological ideology like objects that are produced, managed, optimised and utilised' (Trampe 2001, 238). Finally, ecomodernist discourses present resource-intensive (high-tech) practices as more desirable, even environmentally friendly, while presenting traditional farming as old-fashioned (Asafu-Adjaye et al. 2015). These particular discourses sustain the industrial agriculture practices that deeply disrupt global biosphere integrity and further drive the alienation between humans and the natural environment – they are examples of both growthism and anthropocentrism, which are, according to our ecosophy, *destructive*. For Stibbe, to deal with these *destructive* discourses is to 'raise awareness that the ideology conveyed by such discourses is just a story, and that story has harmful effects' that they are promoting (Stibbe 2015).

Singapore's approach to food security and sustainability has been extolled as a global example to emulate (CNA Insider 2020), which all the more underscores the importance of introducing an eco-critical lens to it. The goal of the ecolinguistics approach within this case study is to encourage individuals (including readers based in Singapore) to be more critical of the claims being made about the purported sustainability of Singapore's agri-food system, and to evaluate the status quo and planning philosophies of this system. Such a critical approach needs to balance the scientific, analytical and technological with ecological and ethical considerations. With reference to our ecosophy, this entails a need to deeply reflect on our moral values of care, connectedness and solidarity with the oppressed. In this case, the oppressed are the non-human lives at stake in Singapore's agri-food approach and that needs to be centred in discourses about sustainability in Singapore. All too often our fellow non-human animals are erased from the stories being told in Singapore, and this *erasure* is present in the agri-food texts of Singapore, too.

Hence, in this case study, I will turn an eco-critical lens onto Singapore's agri-food sector. This section will proceed as follows: first, using a historical analysis, I highlight how the growthism underpinning Singapore's agri-food transition coupled with its contradictions has resulted in a highly particular and unique agri-food landscape. Second, I apply an ecolinguistics lens onto the public communications surrounding Singapore's agri-food industry to reveal the underlying ideologies. An ecolinguistical analysis reveals three dominant discursive strategies: *euphemization strategies* around agrotechnological ambitions that obscure exploitative practices, *appraisal patterns* that privilege modernity over traditional ecological knowledge and *conviction discourses* that justify the control of nature as sustainable. These discourses highlight the growthist and anthropocentric ideologies embedded in Singapore's approach to agri-food, which is deemed *destructive* when evaluated from the ecosophy of this book. Overall, this case study seeks to facilitate a deeper conversation of the ideologies shaping Singapore's agrotechnological ambitions while advocating for a transformative shift towards sustainable practices that honour human–nature harmony and an ethics of care.

From Past to Present: A Historical Analysis of Singapore's Agri-Food Industry

Before we delve into the core discourses within Singapore's agri-food system, it is essential to provide some historical context of agri-food texts in Singapore. These 'texts' are biophysical ecologies of alive places in Singapore, as well as in print and online. To begin, since independence, Singapore's agri-food landscape has undergone three major phases: first, the enhancing of food self-sufficiency (pre-1980s); second, the shift to import-dependent food security (1980s–2000s); and third, the revitalization of local food production (2000s onward to present)

Singapore is often viewed as a developmental city-state that has effectively leveraged land-use planning to enable rapid modernization. Since the 1966 and 1967 Land Acquisition Act, the government has acquired 76.2 per cent of land for 'public purpose', making it the biggest landowner in Singapore by 1985 (Oon and Lim 2014). The term 'public purpose' enabled urban redevelopment as a means to drive economic growth. From a Marxist lens, this ownership of land as the means of production situates the state as the property class. This creates a class relation between the ruling elite and the wider society, whereby the ruling elite has stronger control over development trajectories and the lives of people. This has indeed been the case, with the Urban Redevelopment Authority largely controlling what gets built, by whom, where and for what purposes throughout Singapore (URA n.d.). Sociocultural anthropologist Cynthia Chou's 1989 thesis on the rural sector transition in the 1980s documents how this class relation enabled the ruling elite to impose their ideology through development policies (Chou 1989). Since the ruling elite are positioned to shape land-use policies, and they see their role as such, their planning ideology becomes crucial. Singapore's First Deputy Prime Minister, Mr Goh Chok Tong's speech in 1986 reflects how the ruling elites of that era viewed their role: 'Unless the elite takes on these responsibilities of society, there will be no cultivated society, not even a successful nation' (C. T. Goh 1986). Therefore, the ideologies that exist in the upper echelons of Singapore's government often permeate into how the government fulfils its role as the property owner, the development planner and the regulator. As such, the breadth of possibilities for politico-institutional change post-independence is conditional on the politics of the ruling elite.

It might be difficult for many to imagine Singapore's early independence food story, and even more difficult to condense a complex agricultural history. However, for much of Singapore's early independence, agri-food policy was aimed at increasing self-sufficiency. In 1959, the Primary Production Department

(PPD) was formed to serve the needs of farmers and fishermen. Between 1960 and 1980, the focus of PPD's work was to encourage productivity among farmers to reduce dependence on food imports and improve food security (Gamboa and Chew 1984). As a result, up to the 1980s, Singapore was 95 per cent self-sufficient in pig production and 100 per cent self-sufficient in chicken eggs. At the core of Singapore's food resilience in this era were small-holder semi-subsistence family farms that employed a diversity of mixed-use farming practices, embodied a deep connection to their vocation and stewarded over the land (Chou 1989). Singapore's early farms were an important livelihood and way of life for many and a key pillar of localized food resilience for many more.

The transition in food policy towards import dependency arose in 1984 when Dr Goh Keng Swee was assigned to shape Singapore's food policies at PPD. In his book, *The Economics of Modernization*, he reflects on his developmental ideology: he writes that he, alongside other Asian economists, believed that since resources were limited, planners had to push to cultivate the comparative advantage of a preferred leading sector – for him the ideal sector for Singapore was manufacturing (K. S. Goh 1972, 8). His ideology permeated into land-use policy by reallocating land in a manner that favoured economic growth, with this growth decoupled from subsistence agriculture. He argued that Singapore should produce goods and services that the country would have a comparative advantage in and then buy imported foods with the money earned (K. S. Goh 1984). In the following years, agricultural land in Singapore was systematically reduced until only 1 per cent was designated for agricultural use in 2023 (SFA 2023a). For comparison, 13 per cent is allocated to industry (MND 2019). This effectively spelled the end of Singapore's agri-food self-sufficiency. Furthermore, governmental policy and incentives switched from supporting Singapore's small-holding family farms spread throughout the island at the time of independence, where many lived a more 'traditional' *kampong* lifestyle, where these were to be phased out in favour of food import dependence and the development of industrial agrotechnological show farms. The PPD's focus turned to the management and inspection of imported food products and on developing agrotechnological show farms, to research, develop and sell Singaporean-based agrotechnology that would contribute to the knowledge economy (Chou 2014).

This process of (1) buying out small-holder farms during the early independence era and moving citizens into centrally planned housing flats; and (2) reallocating now cleared land from family-farm and subsistence agriculture to other land uses of higher economic productivity, such as industrial manufacturing, is one of growthism within Singapore's land-use planning, making this switch one in support of a *destructive* ideology and thus discourse and *story-to-live-by*. This deprioritizes activities with essential social reproductive capacities in favour of income-generating activities (Paltasingh and Lingam 2014). Family farms throughout Singapore performed much of the former, as their output was semi-subsistence. Moreso, such farms were organized by kinship relations that are not easily categorized as wageworkers, and these farms performed the care work of feeding their communities and tending to the land. Despite enhancing Singapore's food resilience and self-sufficiency since 1946, they were systematically phased out beginning in 1984.

The consequences of these growthist developmental policies continue through the time of writing. As of 2024, Singapore's food policy is predominantly (90 per cent) reliant on food imports from about 187 countries (Tan 2024). In an era of worsening climate change and limits to global food chains (Mbow et al. 2022), the vulnerability of Singapore's food system became highly visible during the 2008 global food crisis (P. Teng and Escaler 2010) and during the Covid-19 pandemic (P. Teng 2020). After all, the international supply chain of natural gas for fertilizer, and of feedstock and grains, all of which are heavily dependent on oil, is a supply chain that is not as certain and stable as hoped. As a case in point, given the global food industry's heavy reliance on oil, the rise of oil prices in Europe from 2021 to 2022 caused rippling price inflation throughout the supply chain, eventually inflating domestic prices in Singapore (MAS 2022). Coming to terms with these systemic vulnerabilities of the global agri-food system is forcing a rethink of Singapore's food import dependence.

As such, in the third phase of Singapore's agri-food story, the Agri-Food and Veterinary Authority of Singapore (AVA) was set up on 1 April 2000 to increase the local production of vegetables, fish and eggs. More recently, in April 2019, the '30 by 30' goal was announced as a goal to produce 30 per cent of Singapore's nutritional needs locally by 2030. This project even received much critical acclaim in pro-sustainability networks (Hong 2023) and was seen as an example of ecomodern potential (Min 2020). This has been complemented by an acceleration of funding, regulatory changes and government support for the cultivation of a domestic local agri-food industry. The Singapore Food Agency (SFA) would also replace AVA as the lead government agency coordinating such efforts. For instance, the $60 million Agri-Cluster Transformation (ACT) fund has been set up by SFA to support the agri-food sector to become 'highly productive, climate resilient and resource-efficient through capability upgrades, innovation and test-bedding and tech-upscaling' (SFA 2024a). Additionally, since 2022, longer land leases have been tendered to provide more certainty for food businesses (Lim 2022). The masterplan to redevelop the Lim Chu Kang (LCK) region into a high-tech agri-food zone is the most recent and ambitious facet of Singapore's food production story (SFA n.d.). SFA hopes that through the redevelopment, the LCK agri-food zone can potentially produce more than three times its current food production capacity and do so in a sustainable and resource-efficient manner by optimizing for economies of scale (SFA, n.d.). Additionally, knowledge institutions have also supported these efforts. For instance, The Agency for Science, Technology and Research initiated a new food and biotechnology innovation institute to boost Singapore's local nutrition and well-being research and production capabilities (A*STAR 2024). Singaporean Universities and Polytechnics have also developed undergraduate and Workforce Skills Qualifications (WSQ) courses in food technology, agrotechnology and urban farming.

However, issues with high-tech agriculture are emerging due to its structurally high operating costs of production that struggle to compete with more cheaply and conventionally grown imported produce. As a result, many farms in Singapore have been closing down (Loh 2024). This is observed not just in Singapore, but globally, as sustainable food scholar Bomford (2023) points out that fully controlled environments have yet to produce commercially viable examples of staple food production. Anthropologist Vanessa Koh goes a step further to discuss the emergence and systemic failure of agri-tech in Singapore as a product of the politics of 'green capitalism' (Koh 2025; Ponte 2023). She points out that despite their high rate of failure, agri-tech start-ups continue to emerge as they are buoyed by investment funding and state sponsorship premised on the belief in the potential for profitable technocratic solutions to climate change and food insecurity.

Overall, growthism in Singapore's developmental policies has led to a path-dependent process riddled with contradictions that has ultimately led to a complex (and troubled) agri-food industrial landscape present today. Having exceeded its ecological carrying capacity, coupled with its heavy reliance on cheap food imports, the state is hard-pressed to ensure food security for Singaporeans. Additionally, the state plays a dominant hand in shaping socio-political relations and organizing capital to bring about agrotechnological industrial development, and this approach by the state can be deemed as authoritarian high modernism (Scott 1998). This is evident in Singapore's efforts in promoting the research and use of agri-tech, encouraging skill development to secure labour forces, and rationalizing nature into its utility value to industrial production. As a key advocate for the agro-industrial sector, the way the state conducts linguistic representation of nature (and animals) has significant importance: the state is actively creating a *story-to-live-by* for all Singaporeans, with certain ideologies contained within this story. This is especially important because no other industrial sector carries a heavier responsibility for animals. Yet, there has been minimal focus on the discourses surrounding Singapore's 30-by-30 agri-food regime ambitions. In other words, the politico-institutional rules of the agri-food transition have largely remained unquestioned, despite them leading to entrenched routines and practices that favour a highly particular approach to agri-food production.

In the following section, I highlight the cognitive and normative discourses embedded in Singapore's approach to the agri-food system through an ecolinguistical analysis. As will be seen, based on our ecosophy, the following three stories about food that Singaporeans live by are destructive:

1. Agrotechnology as Aspiration
2. Modernity over Tradition
3. Control of Nature

Story #1: Agrotechnology as Aspiration

Agricultural technologies have been evolving in Asia for centuries since settlers began farming in the Golden Crescent. However, the acceleration in the application of technologies to agriculture has been a relatively recent phenomenon with widespread implications for planetary health. The Green Revolution in the early twentieth century saw the introduction of high-yielding varieties of staple crops, particularly wheat and rice, alongside the increased use of chemical fertilizers, pesticides and advanced irrigation techniques. This ability to supply a surplus of cheap grain enabled cities like Singapore to become wholly dependent on rural agricultural land for food at cheap prices.[7] This faith in agrotechnology-empowered food production to address food security persists today. For example, SFA's digital publication *Food for Thought* often highlights the promise of high-tech farming to enhance food security in Singapore (SFA 2022). The reasons for high-tech farming include growing more food on limited land, mitigating climate change and disease and supplying nutritionally enhanced food. Undoubtedly, making nature 'productive' has been the foremost priority for Singapore, with the emphasis on 'production' signalling big economic ambitions – here the story and ideology of 'growthism' is once again central to the government's vision for Singapore.

In SFA's 'Our Vision', a series of photos and illustrations were used to frame the high-tech vision for intensive farming in controlled indoor environments in the Lim Chu Kang Masterplan – a 'high-tech agri-food zone' (SFA n.d.). Artist impressions of model commercial Singaporean farms guide the imagination of the reader to view high-tech farms as possible and ideal. For example, one image in SFA's 'Our Vision' illustrates an abundance of vegetables hydroponically grown along the walls of the corridor, mimicking the customer experience of walking along a supermarket aisle picking out groceries. It also shows the ability to produce fish in a controlled artificial environment, located efficiently on the floor. These illustrations depict the vision for agro-technologies to make efficient the reproduction of non-humans for human consumption. By doing so, they accentuate human–animal relations as one of 'nature-the-commodity and human-the-consumer'. This visual *framing* reifies living beings to be economic-technological objects to be produced, managed, optimized and utilized (Wilhelm Trampe 2017). From the lens of our ecosophy, this is *destructive* as the relation of care between people and grown living beings is determined by an anthropocentric transactional value, and nature is only viewed through its utilitarian value to be produced and sold as food.

But one does not have to look to the future to recognize the human–animal alienation already happening in Singapore's agri-tech capabilities. Descriptions of Singapore's egg-producing factories such as 'hi-tech egg farm', 'produced 433 million eggs', 'modern infrastructure and facilities', and 'cage-free' can be described as a *euphemization strategy* to distance the exploitation processes actually happening in egg production (Wilhelm Trampe 2017). As of May 2025, Singapore has three egg farms, with the fourth experiencing delays and reconsidering the viability of locating operations in Singapore. Common depictions of Singaporean egg-producing factory farms feature densely stacked rows of egg-laying hens in confined spaces, their heads sticking out of narrow slits in the cage grills as a worker in personal protective equipment surveys the setup. Other depictions show hundreds of eggs on an industrial belt system to be processed and packaged. However, this discourse validates and normalizes a distant human–nature relation in the agri-food industry. In these depictions, hens and their egg-laying are *framed* solely as a factor input in an egg production line, rather than birds with full lives. Additionally, eggs are depicted as an industrial output, like cars on an assembly line. Furthermore, these depictions normalize the distant and alienated interaction between humans and the chickens on the farm. It reinforces the dangerous idea that farming is a process of humans exercising control and dominion over the lives of chickens by depriving their natural way of life as creatures and utilizing their biological functions to produce egg-calories for human consumption. Yet, it is seen as 'progress' according to Singapore's ideology of high-productivity agro-tech orientation.

Factory farms are often justified as pragmatic ways for the mass production of food to ensure food security. Such logic is often used to gloss over the politically challenging discourses about the heavy costs of industrial agri-food systems and endeavours to reshape food norms. Such discourses attempt to further validate, justify and conceal the industrialized harm done to animal bodies by framing such harm through the lens of 'high-tech' solutions. Furthermore, technological solutions often further the human–animal alienation process, especially through the use of artificial intelligence (AI) technologies (Teng 2022) and further erodes the relation of care between people and farmed animals (Bossert and Coeckelbergh 2024). Through our ecosophy, however, we reject that food security depends on the violent exclusion of non-humans from our circle of care, nor does it justify their exploitation. Rather, our ecosophy spurs us to rethink farms in a way that foregrounds care and connection between humans and non-humans and the 'Right of All to Thrive'.[8]

Story #2 Modernity over Tradition

Hydroponic setups of artificial lighting, plastic and steel frames and rows upon rows of leafy vegetables stretching to the far depth of the photos, often dominate popular depictions and imaginaries of what future food production could be. These photos showcase the novel farming capacities developed in Singapore, and SFA often presents these photos as success stories of agrotechnological innovation. In *SG50 Special: Singapore's Food Farms*, AVA declares that 'The world can learn from Singapore's success stories. Singapore has made efforts particularly, to maximise productivity from very limited land resources' (AVA 2015). By presenting agrotechnological capabilities with positive expressions such as 'success', 'uplift productivity', 'future of farming', 'science and world-class innovation' and 'protecting our environment', there is an *appraisal and evaluation pattern* that agro-tech-driven farming is modern and therefore ultimately desirable. Simultaneously, AVA also built up a *negative appraisal pattern* of 'traditional farms' by characterizing them as 'inefficient and labourious'. However, there are ecological implications of a prevalent technological-focused cultural evaluation in which the traditional is regarded as bad, and the modern, good, especially when such an evaluation disregards traditional ecological knowledge in favour of agrotechnological promises.

In the *Singapore Certificate Scheme on Good Agricultural Practice for Vegetable Farming*, vegetable farms are encouraged to cultivate in protected and sterile conditions by enacting barriers against insects, birds and animals (SFA 2015). This practice of enacting controlled environments and using monocropping can be observed in the featured agro-tech farms. The loss of species biodiversity increases the vulnerability of the farmed crops, hence constructing a resource-intensive controlled environment becomes a key imperative. However, this runs contrary to traditional ecological knowledge that recognizes species diversity as the source of ecological resilience (Gunderson 2000). For example, Chou (1989) describes the integrated farming practices of family farmers as staggering crops and livestock. This enabled greater inter-species interaction, ecological balance and higher profits. Pairing nitrogen-fixing legumes like long beans and nutrient-demanding crops like *kailan* can promote healthier plant growth while reducing nutrient input. Family farmers would also breed chickens alongside vegetables to allow the chicken manure to fertilize the vegetables (Chou 1989). These integrated farming practices attempt to mimic the interconnectedness and interdependence within ecosystems to enable more abundance. The deep knowledge and practices required in 'traditional farming' compel us to conceptualize traditional farming as a different form of technology – one that can produce food sustainably. Indeed, Ms. Chou's research found that the 'traditional' farmers knew how to 'use the land better than them (authorities)' in order to maintain an integrated farm and produce food for self-sufficiency, income generation and the community's food security. These traditional family farmers and localized food networks kept Singapore's food supply afloat during the Japanese Occupation when vegetable and pig imports fell threefold and food supplies were hard to come by (Teo, Koh and Nathan 2024; Chou 1989). Furthermore, it was the hands-on nature of the work in traditional farming that helped farmers form a 'strong non-alienating bond' with the work, the land and the food produced. Hence, we see that traditional farming often enables an ineffable bond between farmers, the communities they serve and the land they steward. Per our ecosophy, we should be critical of dismissing the traditional in pursuit of the promises that the modern holds.

Story #3 Control of Nature

I end my ecolinguistical analysis of Singapore's governmental vision of its agri-food industry with the story: 'Control of Nature'. Such a story is ironically the same ideology as Singapore's City in a Garden discourse (see Chapter 4). As explained there, the 'Control of Nature' is still a *destructive* discourse, as it contains arguably the most damaging ideology of all: that humans must, and can, control nature. Depictions of fish farms in Singapore range from *kelongs*, which are traditional offshore open-net fish farms to, more frequently, indoor closed containment aquaculture farms. The closed containment aquaculture system (CCAS) controls the aquaculture environment for high-density cultivation of fish. Popular images often depict multiple large circular tanks with high numbers of fish swimming around in clear water. As argued below, the CCAS is justified through a *conviction* discourse to shape the understanding of CCAS as necessary and a sustainable means in people's minds (Stibbe 2015, 129).

In the Singapore Aquaculture Plan launched in November 2024, the SFA recognizes the challenges of rising sea temperatures, ocean acidification, extreme weather events, diseases and harmful algal blooms to aquaculture that will confront Singapore's domestic aquaculture industry (SFA 2024b, 14). By painting the changing sea environment as a 'problem' that will affect the aquaculture system's health, quality and productivity, it argues for the imperative 'solution': controlling the environment. Subsequently, a *facticity pattern* is used to represent CCAS, with very high facticity, as able to mitigate these environmental risks. Publications by SFA assert that CCAS can 'provide a controlled culture environment and protect the stocks from adverse water conditions' (SFA 2023c, 12), and that CCAS also 'reduces exposure to effects of climate change, pollution, and diseases' (SFA 2022). Furthermore, CCAS can 'optimise water use, reduce environmental impacts and reduce biosecurity' (SFA 2024b, 14). However, such discourses are *destructive* because they pointedly argue that Singapore must control nature to preserve aquaculture productivity rather than presenting nature-based alternatives that are in harmony with nature. They also perpetuate the misconception that our relationship with food necessitates a detachment from nature based on humans controlling nature to maximize the productivity of natural resources for human consumption. These are anthropocentric and are in service of growthism and go against our ecosophy of Care, Community and Critical.

The conviction that agrotechnological systems (such as CCAS) can be a sustainable form of production is tenuous at best, as these systems often entail a significant biocapacity footprint. Additionally, the ecological impact of high-tech intensive farming is often understated or not stated at all, and this *erasure* perpetuates a dangerous false promise that environmental limits can be transcended through technology, a central tenet of ecomodernism. The concept of an environmental limit (i.e. ecological carrying capacity or biocapacity) describes the ability of environmental conditions found in a certain area to support the maximum biodiversity and population sizes of native species in that area. When economic activities surpass the ecological carrying capacity of the land, this requires an import of resources and an export of effluents to support such activities. However, such imports of urban metabolism go against strong sustainability. Singapore's national biocapacity is 0.11 gha/person, while ecological footprint is 6.8 gha/person, implying an exceedance of biocapacity by 6,100 per cent (GFN 2023). This is self-evident in Singapore's high import dependence on critical supplies such as food, water and sand. This is also evident in high-productivity food systems. To 'triumph' over the natural limits of the ecological carrying capacity of scarce land and sea space, industrial mono-crop systems expend fossil fuels to control the environment in a way that enables hoped-for high productivity – be it in tanks, vertical hydroponic stacks, or vertical chicken pens. However, this is still a resource-dependent, energy-dense form of generating calories and is largely reliant on imports into Singapore. Yet, the reality is that these ecomodern visions of agri-food resilience are still unsustainable. For example, it takes about eleven calories of fossil fuel to produce one calorie of seafood in Israel – where ninety-one percent of production comes from local indoor high-tech aquaculture (the same systems that Singapore seeks to replicate). Simply put, to produce more than what nature allows, one must take from nature elsewhere. Exceeding environmental limits in Singapore's agri-food system recreates the same dependency on resources that we have to import – natural gas, fertilizers, feedstock, water and even the labour force. Rather than reforming a system of import dependence, agrotechnology-driven high-productivity farming merely transforms dependence into a different form while perpetuating *destructive* ideologies. Ultimately, the concept of ecological carrying capacity presents a cautionary tale that we must work within ecological limits. As such, this is a *destructive* discourse when measured against our ecosophy as it does not recognize the exceedance of ecological limits and perpetuates cross-border extractivism.

The Road Ahead

Overall, this case study problematizes Singapore's approach to agrotechnological farming. Through a historical analysis, we first shed light on the political economy that shaped Singapore's agri-food landscape today, highlighting the state's struggle in ensuring food security, as well as its dominant role in shaping socio-political relations and organizing capital to bring about agrotechnological industrial development. Thereafter, by adopting an ecolinguistical lens, we unpacked the ecological and moral implications of existing agri-food systems. As Singapore strives for a more resilient agri-food landscape, it must confront the ecological limits of its ambitions and foster ideologies and thus discourses that value sustainability, human–nature harmony and ethics of care alongside productivity. For one, Singapore can create a rich and creative space that empowers communities to explore alternative food networks, sustainable foraging, permaculture and agroecological practices. We must also confront the difficult conversations about what we consume. To quote the filmmaker James Cameron, 'You can't be an environmentalist, you can't be an ocean steward without truly walking the walk and you can't walk the walk in the world of the future, the world ahead of us, the world of our children, not eating a plant-based diet'. The path forward necessitates a commitment to rethink the values and ideologies that underpin our relationship with animals, nature and food, ensuring that future developments align with a more compassionate and ecologically sound vision for food security in Singapore.

Conclusion

Lecarl Lim

In a world teetering on socio-ecological collapse, the urgency to adopt an ecologically informed critical discourse analysis to call out the *destructive stories-we-live-by* cannot be overstated. We stand at a pivotal moment where our current trajectory must shift towards one that aligns with ecological and social well-being. This means two things: First, there is a need to encourage a more critical approach to progress by changing the underlying ideologies that underpin *destructive* lifeways. In Singapore, this practically means that we need to challenge the ideologies of growthism and anthropocentrism, especially where and how these occur in the 'texts' by which Singaporeans live. While it is all too easy to undertake a critical reading of other potentially destructive texts like Orchard Road, Marina Bay Sands, Tuas Port, Changi Airport, Sentosa Island, Raffles Place, or other Singaporean places of conspicuous consumption built upon ideologies of growthism and anthropocentrism, in this chapter, we instead chose two less glamorous but nevertheless important discourses and texts to analyse. Our two case studies highlight the too often hidden assumptions about growthism and anthropocentrism that are prevalent even in discourses that on their surface may appear to be moving Singapore into a sustainable, ecomodern future. Rather, as long as Singapore remains wedded to fossil fuels and petroculture, and to deeply entrenched anthropocentrism, both, including in 'hidden' but essential places like Pulau Bukom and domestic conceptions of agri-food resiliency, then the ways Singaporeans live will remain *destructive* when compared to our ecosophy.

Second, we need to foster active praxis and build our Nowtopias (Carlsson and Manning 2010) through feminist degrowth and strong sustainability approaches in this era of climate change. Quoting the environmental writer and activist Rob Hopkins (2008), 'If we wait for the governments, it'll be too little, too late; if we act as individuals, it'll be too little; but if we act as communities, it might just be enough, just in time'. Ecolinguistical analysis is one of the many tools at our disposal, and it holds the potential to reshape the stories we tell ourselves. We can transform ecomodern and anthropocentric stories to those that envision ways of life to ones that thrive within our planetary boundaries – *destructive* discourses do not have to persist.

Before ending this chapter, we leave you with some reflective questions:

- In considering the case studies that highlight *destructive* discourses, do you find yourself in agreement with the assessment that these discourses are indeed *destructive*, based on your personal ecosophy?
- As you reflect on your own lived experience, what other case studies would you consider as *destructive* and how would you apply an ecolinguistical analysis to evaluate it as such? This is an especially important question for those readers in Singapore.
- How has this chapter inspired you to change your way of being, knowing, eating and thinking about sources of energy and living?

Notes

1 Defined by Stibbe (2015, 17) as 'a story that an area of life is unimportant or unworthy of consideration'.

2 Cited from Pg. 61 and our discussion of 'critical' on that page.

3 Defined by Stibbe (2015, 48) as 'the cognitive imposition of a story from one area of life onto another area'.

4 Defined by Stibbe (2015, 129) as 'stories in people's minds about whether a particular description is true, certain, uncertain or false'.

5 Defined by Stibbe (2015, 162) as 'a story in people's minds that an area of life is important or worthy of attention', which can be built through 'vivid and concrete depiction'.

6 In 2023, Singaporeans consumed approximately 2.16 billion hen eggs, 200,900 tons of chickens, 128,300 tons of pigs, 33,500 tons of cows and 126,500 tons of seafood (SFA 2023b).

7 Note: Singapore has always been a rice importer and historically faced threats of rice shortages (Ludher and Paramasilvam 2018, 17).

8 While rethinking and redesigning the entirety of a food system for a constrained nation of almost six million people is clearly beyond the scope of this case study, let alone book, we do point to exemplars in Singapore who align with our ecosophy: livingsoil.asia, groundupinitiative.org, Urban Jungle Folks, Make A Difference (M.A.D) Garden, and cultivatecentral.com.

References

A*STAR. 2024. 'Singapore Institute of Food and Biotechnology Innovation (SIFBI)'. Agency for Science, Technology and Research (A*STAR). 2024. https://www.a-star.edu.sg/sifbi.

Andres, Gabrielle. 2024. 'Shell Shuts Down Oil Processing Unit on Bukom Island to Investigate Suspected Leak'. *The Straits Times*, 27 December 2024. https://www.straitstimes.com/singapore/shell-shuts-down-oil-processing-unit-at-pulau-bukom-to-investigate-suspected-leak.

Asafu-Adjaye, John, et al. 2015. 'An Ecomodernist Manifesto'. https://www.ecomodernism.org/.

AVA. 2015. 'SG50 Special: Singapore's Food Farms – A Story of Then and Now'. *Food for Thought*, 13 March 2015. https://www.sfa.gov.sg/food-for-thought/article/detail/sg50-special-singapore%27s-food-farms---a-story-of-then-and-now.

Basu, Radha. 2005. 'Two Groups Fear New Shell Plant Will Endanger Marine Life'. *The Straits Times*, 4 July 2005. http://www.wildsingapore.com/news/20050708/050704-1.htm.

Bomford, M. 2023. 'More Bytes Per Acre: Do Vertical Farming's Land Sparing Promises Stand on Solid Ground?' *Agriculture and Human Values* 40 (3): 879–95.

Bossert, Leonie N., and Mark Coeckelbergh. 2024. 'From MilkingBots to RoboDolphins: How AI Changes Human-Animal Relations and Enables Alienation towards Animals'. *Humanities and Social Sciences Communications* 11 (1): 920. https://doi.org/10.1057/s41599-024-03441-3.

Bousso, Ron. 2023. 'Exclusive: Shell Cuts Low-Carbon Jobs, Scales Back Hydrogen in Overhaul by CEO'. *Reuters*, 26 October 2023. https://www.reuters.com/sustainability/climate-energy/shell-cuts-low-carbon-jobs-scales-back-hydrogen-ceo-overhaul-sources-2023-10-25/.

Bousso, Ron. 2024. 'Shell's Bet on Gas Boom Takes Shape with String of Deals'. *Reuters*, 12 July 2024. https://www.reuters.com/business/energy/shells-bet-gas-boom-takes-shape-with-string-deals-2024-07-12/.

Buchholz, Katharina. 2024. 'What Are World Leaders Getting Paid?' *Forbes*, 3 September 2024. https://www.forbes.com/sites/katharinabuchholz/2024/09/03/what-are-world-leaders-getting-paid/.

Campbell, William. 1968. 'How a Refinery Enlarged an Island and Will Join It to Another'. *The Straits Times*, 11 June 1968. https://eresources.nlb.gov.sg/newspapers/digitised/article/straitstimes19680611-1.2.73.

Campbell, William. 1971. 'Swamps Are Replaced by Modern Refineries'. *The Straits Times*, 11 October 1971. https://eresources.nlb.gov.sg/newspapers/digitised/article/straitstimes19711011-1.2.155?qt=bukom,%20resettlementandq=bukom%20resettlement.

Carlsson, Chris, and Francesca Manning. 2010. 'Nowtopia: Strategic Exodus?' *Antipode* 42 (4): 924–53. https://doi.org/10.1111/J.1467-8330.2010.00782.X.

Centre for Liveable Cities. 2021. *Jurong Island: Creating a World-Class Energy and Chemicals Hub*. PDF. Epigram. https://www.clc.gov.sg/docs/default-source/urban-systems-studies/uss-jurong-island.pdf?sfvrsn=9de3cad1_5.

Chin, Hui Shan. 2024. 'NEA Advises Public against Swimming Near Several Beaches after Shell's Oil Leak'. *The Straits Times*, 21 October 2024. https://www.straitstimes.com/singapore/no-sign-of-oil-at-sentosa-beaches-waters-near-desalination-plants-after-shell-s-oil-leak-agencies.

Chou, Cynthia. 1989. *Farmers in Singapore: The Rural Sector in Transition*. Singapore: National University of Singapore.

Chou, Cynthia. 2014. 'Agriculture and the End of Farming in Singapore'. In *Nature Contained: Environmental Histories of Singapore*, 216–40. Singapore: NUS Press Pte Ltd. https://muse.jhu.edu/book/30089.

Chou, Loke Ming. 2023. 'Reef Restoration in Singapore'. In *Handbook on Habitat Restoration: General Principles and Case Studies in Singapore*, 316–20. National Parks Board. https://www.nparks.gov.sg/-/media/handbook-on-habitat-restoration/handbook-on-habitat-restoration-chapter-26.ashx.

CIA. n.d. 'Energy Consumption per Capita Comparison – The World Factbook'. *CIA*. Accessed 16 August 2024. https://www.cia.gov/the-world-factbook/field/energy-consumption-per-capita/country-comparison/.

CNA Insider. 2020. 'What It Takes to Keep 5 Million Fed: Inside Singapore's Race to Food Security | Feeding A City – YouTube'. *CNA Insider*. https://www.youtube.com/watch?v=wc9NfAIwJz8andt=1871s.

CNN. 2024. 'Shell Waters Down 2030 Carbon Emissions Target in Latest Fossil Fuel Industry Backslide'. *CNN*, 14 March 2024. https://edition.cnn.com/2024/03/14/climate/shell-carbon-emissions-targets-climate/index.html.

Cornelius-Takahama, Vernon, and Janice Loo. 2016. 'Pulau Bukom'. National Library Board. 2016. Accessed 3 September 2024. https://www.nlb.gov.sg/main/article-detail?cmsuuid=93a29125-66aa-46c6-8ca3-9b9b0f65ab8d.

Cronon, William. 1991. 'The Wealth of Nature: Lumber'. In *Nature's Metropolis: Chicago and the Great West*, 148–206. New York: W.W. Norton.

Daly, Herman E. *Steady-State Economics*. 2nd edn. Washington, DC: Island Press, 1991.

DHI. n.d. 'Reclaiming the Land, Protecting the Environment'. Accessed 12 December 2024. https://www.dhigroup.com/upload/publications/misc/Reclaiming%20the%20land,%20protecting%20the%20environment%20-%20DHI%20Case%20Story%20SG_V1.1.pdf.

EDB. n.d. 'Energy and Chemicals'. Accessed 28 July 2024. https://www.edb.gov.sg/en/our-industries/energy-and-chemicals.html.

EDB. 2024. 'ExxonMobil and Shell Selected to Work with the Government of Singapore on a Carbon Capture and Storage Value Chain'. Press release. 1 March 2024. https://www.edb.gov.sg/en/about-edb/media-releases-publications/exxonmobil-and-shell-selected-to-work-with-singapore-on-a-carbon-capture-and-storage-value-chain.html.

Fang, Andrew. 1961. 'It's a Good, Healthy Life for All on Bukom'. *The Straits Times*, 26 July 1961. https://eresources.nlb.gov.sg/newspapers/digitised/article/straitstimes19610726-1.2.135.19.

Foster, John Bellamy, Brett Clark and Richard York. 2010. 'Capitalism and the Curse of Energy Efficiency'. *Monthly Review*, 1 November 2010. https://monthlyreview.org/2010/11/01/capitalism-and-the-curse-of-energy-efficiency/.

Fu, Grace. 2021. 'Opening Speech at National Energy Efficiency Conference 2021 – Ms Grace Fu'. *Ministry of Sustainability and Environment*, 12 October 2021. https://www.mse.gov.sg/resource-room/category/2021-10-12-speech-at-neec/.

Gamboa, Evangeline, and Lillian Chew. 1984. 'Singapore Sets Its Sights on Feeding Itself'. *Straits Times*, 17 January 1984. https://eresources.nlb.gov.sg/newspapers/digitised/article/straitstimes19840117-1.2.36.

Gan, Kim Yong. 2024. 'Grow Our Economy'. Transcript of speech delivered at MTI's Committee of Supply Debate 2024, 1 March 2024. https://www.mti.gov.sg/-/media/MTI/COS-2024/Speeches/Speech-by-Minister-Gan-Kim-Yong-at-the-MTI-COS-2024.pdf.

GFN. 2023. 'Ecological Deficit/Reserve'. Global Footprint Network. https://data.footprintnetwork.org/?_ga=2.219458586.2062667051.1715067602-275736257.1715067602#/.

Glenn, C. B. 2004. 'Constructing Consumables and Consent: A Critical Analysis of Factory Farm Industry Discourse'. *Journal of Communication Inquiry* 28 (1): 63–81.

Goh, C. T. 1986. *A Developed Economy, A Cultivated Society, A Nation of Excellence*. Singapore: Lecture at The Alumni International Lecture, 1 December 1986. https://www.nas.gov.sg/archivesonline/data/pdfdoc/gct19861201s.pdf

Goh, Keng Swee. 1972. *The Economics of Modernization and Other Essays*. Singapore: Asia Pacific Press.

Goh, Keng Swee. 1984. 'Self-Sufficiency Not the Aim'. *The Straits Times*, 18 March 1984. https://eresources.nlb.gov.sg/newspapers/digitised/article/straitstimes19840318-1.2.53.

Gunderson, L. H. 2000. 'Ecological Resilience – In Theory and Application'. *Annual Review of Ecology and Systematics* 31: 425–39. https://doi.org/10.1146/ANNUREV.ECOLSYS.31.1.425/CITE/REFWORKS.

Harris, Daniel L., et al.2018. 'Coral Reef Structural Complexity Provides Important Coastal Protection from Waves Under Rising Sea Levels'. *Science Advances* 4 (2): eaao4350.

Heng, Swee Keat. 2023. 'DPM Heng Swee Keat at the Launch of the SGX Sustainability Reporting Review'. Singapore, November 23. https://www.pmo.gov.sg/Newsroom/DPM-Heng-Swee-Keat-at-the-Launch-of-the-SGX-Sustainability-Reporting-Review DPM Heng Swee Keat at the Launch of the SGX Sustainability Reporting Review.

Hickel, Jason, and Giorgos Kallis. 2020. 'Is Green Growth Possible?' *New Political Economy* 25 (4): 469–86.

Hong, Alex. 2023. 'Securing Singapore's Future (I/II): The 30 by 30 Food Security Initiative | Illuminem'. Illuminem. Winter 28, 2023. https://illuminem.com/illuminemvoices/securing-singapores-future-the-30-by-30-food-security-initiative.

Hopkins, Rob. 2008. 'Responding to Greer's Thoughts on "Premature Triumphalism"'. *Transition Culture*. https://www.transitionculture.org/2008/11/20/responding-to-greers-thoughts-on-premature-triumphalism/.

IFRC. 2024. 'Early Warning Systems Only Way to Begin Tackling Unprecedented Number of Philippines Typhoons'. Press release, 18 November 2024. https://www.ifrc.org/press-release/early-warning-systems-only-way-begin-tackling-unprecedented-number-philippines.

International Energy Agency. n.d. 'Greenhouse Gas Emissions from Energy'. IEA. Accessed 10 December 2024. https://www.iea.org/data-and-statistics/data-product/greenhouse-gas-emissions-from-energy.

International Energy Agency. 2018. 'The Future of Petrochemicals'. https://www.iea.org/reports/the-future-of-petrochemicals.

Koh, Vanessa. 2025. 'Growing Sustainable Capitalism: The Environmental Politics of Agri-tech in Singapore'. *Environment and Planning A: Economy and Space* 2025: 0308518X251323148.

Lay, Belmont. 2021. 'S'pore Most Tired Country in the World'. Mothership. https://mothership.sg/2021/09/singapore-most-fatigued/.

Lee, Hsien Loong. 2014. 'Transcript of Speech by Prime Minister Lee Hsien Loong at the Opening of Exxonmobil's Chemical Plant Expansion'. *Prime Minister's Office Singapore*, 8 January 2014. https://www.pmo.gov.sg/Newsroom/transcript-speech-prime-minister-lee-hsien-loong-opening-exxonmobils-chemical-plant.

Li, Mei, Gregory Trencher and Jusen Asuka. 2022. 'The Clean Energy Claims of BP, Chevron, ExxonMobil and Shell: A Mismatch Between Discourse, Actions and Investments'. *PLoS ONE* 17 (2): e0263596. https://doi.org/10.1371/journal.pone.0263596.

Lim, Vanessa. 2022. 'More Land and Sea Spaces with Longer Leases to Be Tendered out for Farming to Bolster Singapore's Food Security'. *Channel News Asia*, 7 March 2022.

https://www.channelnewsasia.com/singapore/more-land-sea-spaces-longer-leases-farming-food-security-2544781.

Loh, Renald. 2024. 'With Several Farms Closing or Struggling to Break Even, What Is the Future for Agriculture in Singapore?'. *Channel News Asia*, 19 October 2024. https://www.channelnewsasia.com/today/big-read/high-tech-low-returns-farming-4684566.

Ludher, Elyssa, and Paramasilvam Thinesh Kumar. 2018. *Food and the City: Overcoming Challenges for Food Security*. Singapore: Centre for Liveable Cities Singapore.

MAS. 2022. 'Effects of Higher Global Input Costs on Prices in Singapore'. *Monetary Authority of Singapore*. https://www.mas.gov.sg/-/media/mas/epg/mr/2022/oct/mroct22_sf_a.pdf.

Mbow, Cheikh, et al. 2022. 'Food Security'. In *Climate Change and Land: An IPCC Special Report on Climate Change, Desertification, Land Degradation, Sustainable Land Management, Food Security, and Greenhouse Gas Fluxes in Terrestrial Ecosystems*, 437–550. Cambridge University Press. https://doi.org/10.1017/9781009157988.007.

Menon, Ravi. 2024. 'Opening Remarks by Singapore's Ambassador for Climate Action at COP29 Singapore Pavilion'. *National Climate Change Secretariat*, 12 November 2024. https://www.nccs.gov.sg/opening-remarks-by-singapore-s-ambassador-for-climate-action-at-the-cop29-singapore-pavilion/.

Milstein, Tema, Mariko Oyama Thomas, Jeff Hoffmann and John Carr. 2023. '"Even I Am a Part of Nature": Unraveling the Human/Nature Binary to Enable Systems Change'. *Environmental Communication* 17 (4): 421–36. https://doi.org/10.1080/17524032.2023.2199946.

Min, Ai Kok. 2020. 'Singapore's Emerging AgriTech Ecosystem'. *UNDP*, 16 January 2020. https://www.undp.org/policy-centre/singapore/blog/singapores-emerging-agritech-ecosystem.

Ministry of Sustainability and Environment. 2024. 'Green Economy'. *Singapore Green Plan 2030*. 3 December 2024.

MND. 2019. 'Written Answer by Ministry of National Development on Land Use'. *Ministry of National Development*. 1 April 2019.

Moore, Jason W. 2017. 'The Capitalocene, Part I: On the Nature and Origins of Our Ecological Crisis'. *The Journal of Peasant Studies* 44 (3): 594–630. https://doi.org/10.1080/03066150.2016.1235036.

National Environment Agency. 2024. 'Singapore's Third National Climate Change Study'. https://www.mss-int.sg/docs/default-source/v3_reports/v3-sr-executive-summaryedcdccc564064d97b562d63c04b5436c.pdf.

Neo, Xiaoyun, and Matthew Schneider-Mayerson. 2021. 'Nature, Disappeared: Anti-environmental Values in Singapore's History Textbooks, 1984–2015'. *Environmental Education Research* 28 (1): 56–74.

Ng, Chin-Soon Lionel, and Loke-Ming Chou. 2018. 'Coral Reef Restoration in Singapore – Past, Present and Future'. In *Sustainability Matters: Environmental Management in the Anthropocene*, 3–23. Singapore: World Scientific.

Ng, Weng Hoong. 2013. *Singapore, the Energy Economy*. https://doi.org/10.4324/9780203157732.

Oon, Shaun, and Tin Seng Lim. 2014. 'Land Acquisition Act 1966'. *Singapore Infopedia*. 8 April 2014.

Paltasingh, Tattwamasi, and Lakshmi Lingam. 2014. '"Production" and "Reproduction" in Feminism: Ideas, Perspectives and Concepts'. *IIM Kozhikode Society and Management Review* 3 (1): 45–53. https://doi.org/10.1177/2277975214523665.

Planetary Boundaries Science. 2024. 'Planetary Health Check 2024: A Scientific Assessment of the State of the Planet'. https://www.planetaryhealthcheck.org/storyblok-cdn/f/301438/x/a4efc3f6d5/planetaryhealthcheck2024_report.pdf.

Ponte, S. 2023. Green capitalism and unjust sustainabilities. *Samfundsøkonomen* 2019 (4): 102–8.

r/singapore. 2024. 'R/Singapore on Reddit: Over 100 Quail Chicks Found in Rubbish Bin Were Reportedly from s'pore Farm Quailico Eggs'. *Reddit.Com*. 21 December 2024.

S4F. 2022. 'Fossil-Fuelled Universities: A Call for Universities to End Links with the Fossil Fuel Industry'. https://cdn.prod.website-files.com/660bb195d445f22b836d4636/666bf9b5c02ad1e045d1638f_Fossil%20Fuelled%20Universities%20Report.pdf.

Schneider-Mayerson, Matthew. 2020. 'Introduction: Seeing Singapore with New Eyes'. In *Eating Chilli Crab in the Anthropocene*, 9–18. Singapore: Ethos Books.

Scott, James C. 1998. *How Certain Schemes to Improve the Human Condition Have Failed*. New Haven: Yale University Press.

SFA. 2015. *Good Agricultural Practice for Vegetable Farming (GAP-VF)*. Singapore: SFA.

SFA. 2022. 'Behind the Science of High-Tech Farming'. *Food for Thought*, 28 January 2022. https://www.sfa.gov.sg/food-for-thought/article/detail/behind-the-science-of-high-tech-farming.

SFA. 2023a. 'Our Singapore Food Story'. *Singapore Food Agency*. 2023. https://www.sfa.gov.sg/food-farming/sgfoodstory/our-singapore-food-story.

SFA. 2023b. *Singapore Food Statistics Report 2023*. Singapore: Singapore Food Agency.

SFA. 2023c. *Starting a Sea-Based Farm: An Industry Guide*. 1st ed. Singapore: Singapore Food Agency. https://www.sfa.gov.sg/docs/default-source/food-farming/industry-guides/sfa-sea-farming-guide.pdf.

SFA. 2024a. *Agri-Food Cluster Transformation (ACT) Fund*. Singapore: Singapore Food Agency. 21 October 2024. https://www.sfa.gov.sg/recognition-programmes-grants/grants/agri-food-cluster-transformation-act-fund.

SFA. 2024b. *Singapore Aquaculture Plan*. Singapore: Singapore Food Agency.

SFA. n.d. 'Our Vision'. *Singapore Food Agency*. Accessed 21 December 2024. https://www.ourfoodfuture.gov.sg/master-planning-lck/ourvision/.

Shabana Begum. 2024. 'Over 100 Quail Chicks Found in Rubbish Bin Were Reportedly from S'pore Farm Quailico Eggs'. *The Straits Times*, 11 November 2024.

Shell. n.d. 'Our Company History'. Accessed 29 August 2024. https://www.shell.com/who-we-are/our-history/our-company-history.html.

Shell. 2010. 'Shell Completes Its Largest Petrochemicals Project'. 4 May 2010. https://www.shell.com/business-customers/chemicals/media-releases/2010-media-releases/shell-completes-its-largest-petrochemicals-project.html.

The Shell Company of Singapore Limited. n.d. *Shell Pulo Bukom*. Singapore: Public Relations Department, The Shell Company of Singapore Limited. From the Lee Kip Lee and Lee Kip Lin family archives: Shell and Pulau Bukom publications (Call number: RCLOS 338.2728095957).

Shell Global. n.d. 'Greenhouse Gas Emissions'. Accessed 2 December 2024. https://www.shell.com/sustainability/transparency-and-sustainability-reporting/performance-data/greenhouse-gas-emissions.html.

Shell Global. 2010. 'Shell Completes Its Largest Petrochemicals Project'. Shell Global. Accessed 4 May 2010. https://www.shell.com/business-customers/chemicals/media-releases/2010-media-releases/shell-completes-its-largest-petrochemicals-project.html.

Shell Singapore. n.d. 'Shell Energy and Chemicals Park Singapore'. *Shell*. https://www.shell.com.sg/about-us/projects-and-sites/pulau-bukom-manufacturing-site.html.

Shell Singapore. 1976. *Shell in Singapore*. From the Lee Kip Lee and Lee Kip Lin family archives: Shell and Pulau Bukom publications (Call number: RCLOS 338.2728095957).

Shell Singapore. 2019. 'Shell Completes Storage Expansion at Singapore Refinery'. Press release. Accessed 6 June 2019. https://www.shell.com.sg/media/2019-media-releases/shell-completes-storage-expansion-at-singapore-refinery.html.

Shell Singapore. 2024. 'ExxonMobil and Shell Selected to Work with the Government of Singapore on a Carbon Capture and Storage Value Chain'. Press release. Accessed 1 March 2024. https://www.shell.com.sg/media/2024-media-releases/exxonmobil-shell-work-with-singapore-government-on-carbon-capture-and-storage-value-chain.html.

Singapore Parliament. 2005. 'Oral Answers to Questions'. *Parliament 10, Session 2*, vol. 80, sitting 8 (20 July 2005). 'Land Reclamation At Pulau Ular (Environmental impact)'. Dr Geh Min, Mr Heng Chee How, Mr Abdullah Tarmugi.

Singapore Parliament. 2004. 'Written Answers to Questions for Oral Answer Not Answered by End of Question Time.' *Parliament 14, Session 2*, vol. 95, sitting 138 (6 August 2024). 'Greenhouse Gas Emissions Attributable to Pulau Bukom and Jurong Island, and Projected Reduction from Sale of Shell's Oil and Chemicals Facilities'. *Mr Gan Kim Yong, Mr Dennis Tan Lip Fong.*

Singapore Trade and Industry. 1966. From the Lee Kip Lee and Lee Kip Lin family archives: Shell and Pulau Bukom publications (Call number: RCLOS 338.2728095957).

Stibbe, Arran. 2015. *Ecolinguistics: Language, Ecology and the Stories We Live By*. New York: Routledge.

The Straits Times. 1959a. 'Plight of Islanders'. *The Straits Times*, 23 July 1959. https://eresources.nlb.gov.sg/newspapers/digitised/article/straitstimes19590723-1.2.38.

The Straits Times. 1959b. '$40m. Shell Refinery for Singapore'. *The Straits Times*, 1 December 1959. https://eresources.nlb.gov.sg/newspapers/digitised/article/straitstimes19591201-1.2.72.

The Straits Times. 1974. 'Chemical Complex for Bukom under Big Expansion Plan'. *The Straits Times*, 24 October 1974. https://eresources.nlb.gov.sg/newspapers/digitised/article/straitstimes19741024-1.2.46.3.

The Straits Times. 1982. 'Five Isles Off West Coast Will Be Amalgamated into One'. *The Straits Times*, 16 September 1982. https://eresources.nlb.gov.sg/newspapers/digitised/article/straitstimes19820916-1.2.142?qt=pulau,%20busing,%20bukom,%20shellandq=Pulau%20Busing%20bukom%20shell.

Subhani, Ovais. 2024. 'Shell Sells Singapore Assets on Pulau Bukom and Jurong Island after Strategic Review'. *The Straits Times*, 9 May 2024. https://www.straitstimes.com/business/shell-sells-singapore-assets-at-pulau-bukom-jurong-island-after-strategic-review.

Tamisari, Faustina. 2017. 'The Role of State in Economic Development and Industrialisation in Singapore with Export-Led Growth Model'. *International Journal of African and Asian Studies* 30: 81–7.

Tan, Cheryl. 2024. 'S'pore Vegetable, Seafood Production Fell in 2023 Due to Construction Challenges, Inflation: SFA'. *The Straits Times*, 20 May 2024.

Teng, Paul. 2020. 'Assuring Food Security in Singapore, a Small Island State Facing COVID-19'. *Food Security* 12 (4): 801–4. https://doi.org/10.1007/s12571-020-01077-0.

Teng, Paul, and M. Escaler. 2010. 'The Case for Urban Food Security. A Singapore Perspective'. *NTS Perspectives*, January, 1–18.

Teng, Rachel. 2022. 'AI Is Helping Singapore Achieve Food Security. Here's How'. *Govinsider.Asia*, 9 December 2022. https://govinsider.asia/intl-en/article/ai-is-helping-singapore-achieve-food-security-heres-how.

Teo, Shan Chian, Jeffrey Tzu Kwan Valino Koh and Russell Nathan. 2024. 'Examining the Resourcefulness of War Time Recipes to Develop Future Urban Food Systems for Singapore'. In *Diets and Local Food Systems in Asia*. Singapore: Singapore Institute of Technology. https://doi.org/10.25447/SIT.27092437.V2.

Thompson, Andrea. 2024. 'Earth Will Exceed 1.5 Degrees Celsius of Warming This Year'. *Scientific American*, 5 November 2024. https://www.scientificamerican.com/article/2024-will-be-the-first-year-to-exceed-the-1-5-degree-celsius-warming/.

Tilsted, Joachim Peter, Alice Mah, Tobias Dan Nielsen, Guy Finkill and Fredric Bauer. 2022. 'Petrochemical Transition Narratives: Selling Fossil Fuel Solutions in a Decarbonizing World'. *Energy Research and Social Science* 94 (2022): 102880.

Trampe, W. 2001. 'Language and Ecological Crisis: Extracts from a Dictionary of Industrial Agriculture'. In *The Ecolinguistics Reader: Language, Ecology and Environment*, 232–40. London: Continuum.

Trampe, Wilhelm. 2017. 'Euphemisms for Killing Animals and for Other Forms of Their Use'. In *The Routledge Handbook of Ecolinguistics*, 325–41. New York: Routledge. https://doi.org/10.4324/9781315687391-22.

UBS. 2024. *Global Wealth Report 2024*. https://www.ubs.com/global/en/wealthmanagement/insights/global-wealth-report.html

URA. n.d. 'What We Do'. Urban Redevelopment Authority. Accessed 3 January 2025. https://www.ura.gov.sg/Corporate/About-Us/What-We-Do.

Wai, Ronnie. 1982. 'JTC Plans Oil Terminal on Tiny Island'. *The Straits Times*, 19 June 1982. https://eresources.nlb.gov.sg/newspapers/digitised/article/straitstimes19820619-1.2.31.

Ward, James D., Paul C. Sutton, Adrian D. Werner, Robert Costanza, Steve H. Mohr and Craig T. Simmons. 2016. 'Is Decoupling GDP Growth from Environmental Impact Possible?' PloS One 11 (10): e0164733.

Wilson, Sheena, Adam Carlson and Imre Szeman, eds. 2017. *Petrocultures: Oil, Politics, Culture*. London: McGill-Queen's University Press.

Wong, Desmond. 2010. 'Shell Opens Jurong plant'. *Today*, 5 May 2010. https://eresources.nlb.gov.sg/newspapers/digitised/article/today20100505-1.2.5.

Wong, Lawrence. 2022. 'DPM Lawrence Wong at the Singapore International Energy Week'. Prime Minister's Office Singapore, 25 October 2022. https://www.pmo.gov.sg/Newsroom/DPM-Wong-at-Singapore-International-Energy-Week.

4

Ambivalent Discourses of Singapore

Anjali Venketram, Hannah Ng Shueh Yi,
Kristina Gweneth Ponce Simundo and Tan Yock Han

Introduction

Anjali Venketram

Ambivalent discourses reflect the complex and often contradictory narratives that shape Singapore's relationship with the environment. These discourses exist in the grey area where stories people live by can be seen from multiple perspectives and in different lights. What is *beneficial* to one community can be seen as *destructive* to another, and what can seem *beneficial* in the short run may be *destructive* in the long run. As our ecosophy is relatively broad, some aspects of the case studies in this chapter may have discourses that align with our pillars of community, care and critical. However, some discourses, upon closer look, actually go against the goals of our ecosophy in the long term. Hence, the 'texts' analysed in this chapter per Stibbe's definition qualify as *ambivalent* – they reflect the tension and contradictions that arise with differentiated priorities and the challenges of balancing social, environmental and economic needs and priorities, especially from a strong sustainability perspective. Unfortunately, there are no silver-bullet solutions to the climate change crisis and nuances of economic development, gender inequality, racial inequity, species protection and ecomanagerialism, including in Singapore. However, despite these very real challenges, there are many texts in Singapore that are *ambivalent*, and we find it important to point out the *beneficial* aspects of these, while being honest about their *destructive* aspects, too, in the hopes of identifying them as glimpses into the ideal world for which our ecosophy advocates. In our ecosophy, we emphasize the need to be critical and find that the practice of such ecosophically informed constructive criticism involves questioning all environmental narratives while seeking to find compromises that allow us (and the planet) to thrive.

Ambivalent discourses often arise in situations of transition, where societies are rapidly developing, and they reveal tensions between growth and regenerative sustainability, consumerism and conservation and short-term economic gains and long-term environmental well-being. In the field of ecolinguistics, *ambivalent* discourses are important because they reveal that no matter how strongly we stand by our own ecosophies, there arise situations where we have conflicting emotions, choices, values and ideologies and are therefore unable to continuously prioritize strongly sustainable actions and behaviours. This chapter's examination of pre-selected *ambivalent discourses* in Singapore allows us to explore the areas where sustainability efforts in Singapore are evolving; where progress is being made but is often incomplete, and where competing values are in play. A leading tension in Singapore's *ambivalent* discourses often emerges between the ideology of growthism and economic progress that results in infrastructural expansion, and with approaches (from anthropocentric to ecocentric) to ecological preservation and conservation. As covered in Chapter 3's examination of *destructive* discourses, economic progress has historically taken priority over the well-being of the human and non-human residents of Singapore, and this reality means that most of Singapore's ecomodern sustainability efforts to-date will be at best *ambivalent*.

Singapore has long been celebrated locally and internationally as a sustainable and urban environmental planning model. As a 'City in Nature', it has received widespread praise for its green initiatives and its commitment to becoming a leader in smart, sustainable urban design. These accolades are frequently highlighted in global sustainability rankings, and Singapore is often positioned as a paradigm of how cities can thrive while incorporating environmental considerations (Tan 2018). For example, the 2021 Global Sustainability Index recognized Singapore for its commitment to environmental policy and green spaces, pointing to its reputation as a leader in urban sustainability.

While Singapore's environmental policies are lauded for their innovation and effectiveness, they are frequently framed within the broader context of its economic agenda. The city-state's drive for urbanization and technological advancement is frequently justified through the lens of economic competitiveness, with sustainability often framed as a tool to support economic success rather than an end in itself. As *framed* by the Centre for Liveable Cities in Singapore, 'over time, the balancing act [between economic growth and sustainability] has turned into an economic advantage and a virtuous cycle'. The amplification of narratives like 'City in Nature' leads to a deceptively positive understanding of Singapore's struggles with land use and can mask the broader environmental costs of development, thus driving our ecosophically informed urgency to analyse how these issues are framed within the Singaporean context. The decision-makers in Singapore have often emphasized the need to create economic growth to future-proof the nation and to maintain its economic resilience. While the 'City in Nature' aspect of the Singapore story is meant to inspire confidence in the country's ability to meet its environmental (and overall sustainability) challenges, it subtly reinforces the narrative that economic progress must come first and that pathways to sustainability should only be created when they align with economic growth.

In this chapter, we will see how although some narratives and their ideologies are not overtly harmful based on our ecosophy, they can nonetheless inadvertently reinforce 'growth at all cost' systems that prioritize economic growth over the well-being of communities. While these discourses have a well-intentioned aim of dealing with some of the ecological problems caused by *destructive* discourses, they arise from the same society as the *destructive* discourses and can be influenced by political or commercial interests (Stibbe 2015, 29). We offer here an ecolinguistical analysis of the Botanic Gardens and the story of Singapore as 'City in Nature' as two *ambivalent* case studies that show how human-centred power structures are embedded into all of Singapore's environmental decision-making, illuminating how some discourses that seem *beneficial* from a cursory glance are influenced by a *destructive* underlying ideological system. We chose these two examples because they are presented as pinnacles of sustainability in Singapore. The Singapore Botanic Gardens, a UNESCO (United Nations Educational, Scientific and Cultural Organization) World Heritage Site, is often celebrated as a symbol of Singapore's commitment to conservation, biodiversity and public green spaces. It is a place where visitors can connect with nature and experience a vision of ecological harmony within the city, where such spaces are limited. The Gardens are renowned for its orchid varieties and scientific advancements in the fields of botany and ecology. Discussing the 'City in Nature' narrative in a chapter on *ambivalent* discourses is also crucial because it exemplifies the tension between ecological aspirations and economic imperatives. Both these examples embody the ideals of environmental stewardship, but they are deeply intertwined with narratives (present and historical) that prioritize economic progress, global competitiveness and urban development.

City in Nature: Vision or Veneer?

Hannah Ng Shueh Yi, Kristina Gweneth Ponce Simundo, Tan Yock Han

> We will create a green, liveable and sustainable home for Singaporeans… To do so, Singapore will be transformed into a City in Nature. We will build upon what has been achieved and further restore nature into the urban environment. (Ministry of Sustainability and Environment 2024)

This case study of Singapore as a 'City in Nature' is unique amongst the backdrop of this book's ecolinguistical analysis of Singapore's biocultural textual spaces, in that this specific case study is of an abstract yet important, publicly visible, governmentally generated story by which Singaporeans live (or at least, the story by which Singapore's government wants Singapore as a country to live). 'City in Nature' is a significant metaphor that has evolved to become a prominent ideological vision that influences land-use planning in the island-city. As per Stibbe, 'Metaphors use a frame from a specific, concrete and imaginable area of life to structure how a clearly distinct area of life is conceptualised' (2015, 64). 'City in Nature' as a *metaphor* takes the concepts of Nature and inserts Singapore into the wider natural environment, which will be explored here later.

'City in Nature' presents an image of ecotopia, an imagined perfect balance and harmony between urban and natural forces. It is a point of pride in the Singapore Story that the city-state has achieved optimal urban planning and has appropriately demarcated each corner of the country for a decided purpose to fulfil this vision. The City in Nature vision positions Singapore as an exemplary model of balancing the presence of the urban and the natural in a shared space, which is a *story-we-live-by.* This *framing* uses nature, 'a story from one area of life', to conceptualize the city (Stibbe 2015, 47). Nature is then used to structure the concept of Singapore as a city, and hence, it is important for us to critically analyse this story for its potentials and failings as a metaphor that tries to move Singapore onwards (***Majulah…?***) into a hoped-for just and regenerative future.

In this case study, we build on Chapter 1 and use our ecosophy to dive deeper into the ecolinguistical foundations that underlie the City in Nature vision. This vision manifests in specific urban choices made in the land-use planning of Singapore, including in various national landmarks upheld by the urban planning department (the Urban Redevelopment Authority [URA] of Singapore) as tenets of the City in Nature strategy. Some questions that emerge from our analysis of the City in Nature story, informed by using Stibbe's tools of *framing* and *salience*, include: what *stories-to-live-by* may emerge from the City in Nature discourse? Do these stories align with the hopes of our ecosophy? As such, this case study is an exploration of how ecotopic representations of Singapore may appear to be overtly *beneficial* upon cursory glance, but once we probe deeper, we navigate tensions of ecomodernity, strong sustainability and climate change that will have significant implications for various possible futures of urban living in Singapore.

From Garden City to City in a Garden: Origins of the City in Nature story

Before what we now know as the vision of the City in Nature, Singapore was first introduced to the vision of the Garden City in 1967, the original point in the genealogy of the City in Nature. Singapore's Prime Minister at the time, Lee Kuan Yew, introduced it as a vision crucial to the nation's formation of its identity, especially in the early years of the country's independence after colonial rule (see Chapter 1). The Garden City moniker served as a foundational pillar for the burgeoning identity of other infrastructures of Singapore, like its cleanliness, ideas of well-being, tourist appeal and its foreign perception and reputation.

The Garden City is captivating as a developmental allegory – the idea that the natural and urban can exist in perfect, balanced harmony and an alluring image – the picture-perfect intertwining of both elements of a 'forward' nation. However, this vision promotes several underlying discourses, especially that Nature is predominantly 'green' or 'garden'-like. The *Green=Nature=Gardens* discourse promotes two perceptions of nature. First, that nature is green, and second, that nature (at least in Singapore) exists primarily in Gardens.

The first perception creates a hyper-focus on green-ness (e.g. every street in the city being lined with trees, and a reputation of Singapore's iconic green landmarks such as the Botanic Gardens, Gardens by the Bay and its numerous neighbourhood parks). The resulting *erasure* of non-green spaces such as mudflats, intertidal areas, coral reefs and anything underwater from this hyper-focus on gardens becomes problematic because the non-human spaces of Singapore that are not gardens are erased from the overemphasis on green=garden spaces. These *erased* spaces become what we call 'shadow spaces', 'places that provide our material and ecological support' but that are often erased and forgotten (Plumwood 2008).

The second perception of nature contained in the ideological discourse and 'text' of the Garden City is one that positions nature as a resource for anthropocentric use, including for research, recreation and management (see the next case study in this chapter on the Botanic Gardens that elabourates on the harms concealed by garden culture in Singapore.). In most discourses, the term 'Gardens'[1] typically implies they are the product of human manipulation, desires, aesthetics and ideas (Montefrio, Lee and Lim 2021), and this meaning holds for Singapore.

The next significant iteration of the state-driven ideology of nature-as-garden was the City in a Garden, with this branding of Singapore unveiled in 1998. The National Parks Board (NParks), a government body created in 1990, was tasked to uphold the vision of Singapore as a garden city and to help progress Singapore towards more 'biophilic' attempts at restoring habitats within Singapore's ecosystem while involving the community (National Parks Board 2021). This reiteration shifted the image from a city *with* gardens to a city nestled *within* a large garden (National Parks Board 2021). This shifts towards an ecocentric framing, which is a discourse that aligns with our ecosophy and our advocacy for discourses that intertwine the human with the non-human, is helpful. However, it still invokes the image of the 'garden', an image that connotes the hyper-curation of nature and that justifies Singapore's ecomanagerialism. These ideologies manifest in the planting and landscaping methodologies of Singapore – the same ones that lead to the accolades of Singapore being an ecomodern city of the future as shared in the introduction and Chapter 1.

This most recent shift to City in Nature occurs as part of the Singapore Green Plan 2030 (SG Green Plan) (Ministry of Sustainability and Environment 2024). Going further beyond the previously successful Garden City (1967) turned City in a Garden (2012) vision, the SG Green Plan began framing policies and narratives under its *City in Nature* pillar. Linguistically, this shift in nomenclature reflects a movement toward a more holistic inclusion of the natural world. However, central to this discourse is a separation of human and nature and a separation between the entities of urban ('City') and natural ('Nature'/'Garden'). Thus, this state-created vision ideologically divides the human and non-human. This dualism is a key source of tension that makes the City in Nature vision an *ambivalent* discourse per our collective ecosophy.

However, it is evident that across all iterations of the City in Nature vision, the language of separation has become challenged. This is seen in the most recent iteration where 'City' is the noun used to represent Singapore. The word connotes an ideologically urban, anthropocentric lifestyle that reinforces the lived reality of most Singaporeans. However, the preposition 'in' is the crux of the City in Nature discourse in its current iteration, implying the insertion and possibly integration of Singapore's anthropocentric, urban 'city' into the larger natural environment, represented by 'Nature'. By removing 'garden' from it, the discourse has shifted away from a more dominating, anthropocentric view of nature and towards a more inclusive one, where the implication is that the natural world includes Singapore-the-City. The departure from the initial *framing reframes* the discourse from a more *destructive* discourse to one that is currently *ambivalent*. It gives the city more potential to be included into the natural environment, different from the globally dominant dualistic view of urban spaces and nature spaces.

The evolution of the Garden City vision into the City in Nature aligns with certain parts of our ecosophy, including the integration of the urban and the natural, and the human and the non-human, into a joint *Community*. At least from our initial ecolinguistical analysis of the evolution of City in Nature nomenclature, some *beneficial* discourses prevail, such as the 300 additional hectares of naturalized gardens and parks, creating naturalized infrastructure to support biodiversity and reduce flooding risk and species recovery plans for one hundred floral and sixty faunal species.

We move now from analysing the historical discursive context behind the vision to analysing how the concept of the City in Nature manifests in other texts in Singapore. We do this by analysing (1) policies that relate to the City in Nature vision, (2) case studies identified as hallmarks of the vision and (3) implications on spatial planning according to the vision.

'City in Nature' as Policy: Destroying to Create

NParks is responsible for spearheading and achieving the goals of the City in Nature. Apart from the 7,800 hectares of nature reserves, gardens and parks linked by 370 kilometres of park connectors (National Parks Board 2023b), NParks is also responsible for managing five other sub-pillar goals:[2]

1. To grow Nature Park Networks by creating an additional 200 hectares[3] for ecological connection near four legislated nature reserves and for recreational use;
2. To naturalize gardens and parks by naturalizing existing infrastructure and conduct 160 floral and faunal restoration programmes;
3. To restore nature in urban areas by creating 200 hectares of skyrise greenery and planting at least 170,000 trees in industrial areas;
4. To connect green spaces with the goal that every household in Singapore is to be within a ten-minute walk from a park by expanding park connectors

and Nature Ways, and where every road is to be a Nature Way, totalling 300 kilometres of Nature Ways and 500 kilometres of park connectors;

5. Enhancing vet care and animal management through increasing and strengthening capacities for animal health, welfare and management.

The overall net increase in greenery, green spaces and green infrastructure is a policy effort by the state to bring nature to Singapore's cityscape. The sub-pillars directly expand flora and faunal habitat and intentionally include more non-human nature in urban Singapore. The City in Nature policy undoubtedly systemically expands the aspects of the *Care* and *Community* parts of our ecosophy through their sub-pillars aimed at connecting residents with nature and caring for nature through restoration efforts. Certain sub-pillars are also equally intended for human use and benefit for either recreational or infrastructural uses; for example, naturalizing canals for increased flood resilience and greening urban areas provide co-benefits like improved cooling, air quality and mental well-being for human (and non-human) inhabitants of Singapore.

However, while these policy interventions appear to be biophilic and beneficial for the human and non-human, their efficacy is contested by other land-use policy decisions by the state that come at the cost of ecosystem health as a whole. According to a study from Eco-Business, a leading journalism platform for climate issues in Southeast Asia, up to 7,331 hectares of secondary forest may be cleared for redevelopment in Singapore over the next two decades (Hicks 2023). This is no small feat – it is comparable to the 7,800 hectares NParks are responsible for renaturing.[4] Approximately half of the cleared land will be converted into reserve sites, land that the government can sell to developers, and a further 8.8 per cent of the land is slated for developments of parks (Wu 2023). Although secondary forests generally have less ecological value to overall biodiversity and as ecological habitat compared to primary forests, the discourse surrounding secondary forests regularly dismisses their inherent value as non-human living things through their continued and planned deforestation. Ironically, natural ecologies are being cleared only for them to be renatured after. This *erases* their intrinsic value through a discourse that prioritizes urban development, a discourse that dominates Singapore. This *destructive* story continues to prioritize and require the clearing of forests– both primary and secondary, but especially the latter – to make way for new urban developments. A recent well-known example is the Dover Forest. A young patch of secondary forest, it was initially slated for full clearance, but nature enthusiasts swayed the decision and slated half for conservation and half for clearance. This is a prime *ambivalent* example, where there are *beneficial* outcomes (such as the conservation of half of the forest) and equally *destructive* outcomes (the continued clearance of the other half of the forest). Figure 4.1 shows the visual impacts of the clearance from the nearby Mass Rapid Transit station, Dover MRT.

Despite its *destructive* aspects, the policies for City in Nature ironically still *foreground* net-positive nature initiatives, keeping to the same discourse of greenery, combining this with the clean[5] and manicured environments of Singapore. With every additional natural feature of the City in Nature NParks goals, the *ambivalent* tension emerges where on one level Singapore through its ecomodern ecomanagerialism shapes the actual biophysical reality of Singapore as a part of the earth; yet, the accessibility to nature spaces and the design of flora and fauna are *framed* as an addition to Singaporean's anthropocentric lives. Here, the human realm of Singapore's urban residents is still framed as the primary socio-spatial priority of Singapore's land-use planning, and the non-human aspects of Singapore's ecosystems are *framed* as secondary. Thus, this flips Singapore's governmental policies to, at best, weak sustainability. The City in Nature vision from the get-go assumes that urbanization is exclusive of nature and that greening (from the original planting of trees by Lee Kuan Yew to the planned expansions of the 2020s and beyond) is a hallmark of an orientation towards sustainability. It destroys any assumption that people may have of their inherent connection with nature, instead *re-minding* (Plumwood 2008) people that Nature is 'additional', perpetuating a reductive and exclusionary positioning of nature in physical and ideological terms.

Despite the pursuit of creating more spaces of nature and expanding ecological and social connection, the underlying discourse of the City in Nature policy

Figure 4.1 Dover MRT. Left, border between conserved forest area and start of cleared area (now a construction site). Right, cleared land with cranes beginning foundational works. (Photos by Yock.)

continues to espouse an *ambivalent* discourse: *destroying to recreate.* Given the current cycle of development and naturalization, it is prudent to question what types of biodiversity will and will not be preserved under the auspices of NParks. Recovery programmes, habitat restoration and naturalization of urban spaces cannot make up for losing historical, cultural and heritage diversity, and moreso, cannot make up for losing even more non-human habitat. These are in direct opposition to our *Care* and *Community* pillars, as they either do not recognize the inherent value of these spaces, therefore degrading relationships of *Community*, or they recognize their value and intentionally, systemically and continually choose to destroy these non-human spaces. Even the language of 'losing' these spaces is misleading – in actuality, we are choosing development and destruction over maintaining these spaces. Although NParks' plans summarized earlier are positive steps in mitigation and reparations, these are still not silver-bullet technological solutions that can fully make up for the losses of non-human nature that continue to occur. Such loss depletes Singapore's overall resilience, especially in biodiversity health and ecosystem functionality.

Therefore, the City in Nature policy is a prime example of an overarching *ambivalent* discourse rife with tensions and contradictory implications, and this underlying ideology has manifested into land-use policies that inform Singapore's *stories-we-live-by.* In the following sections, we move from an analysis of governmental documents to identifying discourses latent within several biocultural spaces that are 'texts' which offer rich insights into the *ambivalent* discourses within the City in Nature vision of Singapore.

Eco-*Link@BKE*: The Human–Nature Divide

As discussed above, although the use of the preposition 'in' attempts to integrate the two entities of 'City' and 'Nature', we see that in practice, it does not erode the alienation and abstracting of nature from the human. Rather, the 'in' ideologically via language/discourse perpetuates the ideology of anthropocentrism in Singapore. To see how the discursive binary of human/nature that is embedded in the City in Nature vision manifests in different biocultural textual spaces within Singapore, we examine Singapore's Eco-Link@BKE. This ecological bridge sits across one of the major expressways in Singapore, the ten-lane Bukit Timah Expressway (BKE). The Eco-Link is presented as an infrastructural innovation in Singapore's roadways that allows fauna residing in the nature reserves that surround the expressway (Bukit Timah Nature Reserve and Central Catchment Nature Reserve) to cross safely and thus minimizes their vulnerability to roadkill accidents (National Parks Board 2023a). For historical context, the two nature reserves used to be connected before the BKE was built in 1986. The Eco-Link was created in 2013 to circumvent some harms on biodiversity that the creation of the BKE resulted in Figure 4.2.

Our analysis of the Eco-Link @ BKE finds that it upholds the tenets of the City in Nature vision and its desire to find a compromise between preserving nature and developing the urban landscape of Singapore – it is literally an *ambivalent* 'text'. On one hand, the development of the BKE has increased the connectivity of roadways across the island, privileging human mobility and petroculture; however, the Eco-Link has equally reconnected, protected and conserved biodiversity and faunal habitats, especially for species that reside around the roadway.

Upon further inspection, this case study uncovers more *ambivalent* discourses that underlie the City in Nature vision – specifically, that of a permanent hierarchy that centres and *frames* the human over the non-human. The Eco-Link's objective is to separate the biodiversity of these natural reserves from the BKE, which is populated chiefly by motor vehicles and residents in Singapore transiting

Figure 4.2 View of the Eco-Link from the Bukit Timah Expressway. (Photo credit: Hannah.)

throughout the city. Upon first impression, this might seem like something that promotes conservation among Singapore's urban values. However, discursively the elevation of the linkbridge above the highway promotes a discourse of separation – which brings to mind the *framing* of humans as inherently harmful to spaces of biodiversity, necessitating physical division between the humans and non-humans. It presents the BKE highway as a violent intrusion of natural space, having cut through the two previously connected reserves for a human goal – greater connectivity for transport purposes in service of petroculture. While this is true of the multi-lane highway, it has the effect of reifying underlying *destructive frames.* Instead of challenging the behaviour upheld by the BKE, such as its facilitation of high-speed movement through the city via motor vehicles that use combustion engines that release greenhouse gases, the Eco-Link reifies this violence by continuing to uphold the *frame* of necessitated separation from the human and non-human. Such structural innovations become less *beneficial,* as they reinforce petroculture values without addressing the wider implications of harm that are produced from this original road infrastructure development.

While the functional goal of protecting flora and fauna via the Eco-Link is *beneficial* according to our ecosophy's pillar of *Community* that extends societal well-being to include the non-human, parts of the Eco-Link via its *framing* and story is *destructive* according to our ecosophy's goals of *Care.* Our ecosophy advocates for the harmony and healing of the ecosystem through being *Critical* in interrogating the binaries that exist in Singapore within our day-to-day interactions with nature. If we are truly operating with *Care* at the forefront of our relationship with the ecosystems and ecologies of alive places in Singapore, then we would radically transform our ways of navigating space instead of building technological innovations to manage the symptoms of anthropocentric destruction from developments of vehicular infrastructure. A discourse that arises from this phenomenon is that the creation of separate spaces – the Eco-Link for 'nature's' use, and BKE for 'human' use – is necessary. The Eco-Link upholds and reinforces the reinforces the binary discourse. The 'text' of this built part of Singapore's urban ecology reinforces the ideology that nature in Singapore is alienated from the human, instead of integrated in a shared lived reality.

Although on the surface the goal of the Eco-Link may be *beneficial,* nature is nonetheless relegated to the position of a subject to be protected through paternalistic delegations of separate spaces, rather than experienced and integrated with, creating a distant relationship between human and nature. Such ecomanagerial privileging of the human and framing of nature as an object of use for the development of Singapore has been present since Singapore's founding.[6]

Furthermore, this binary discourse does not build resilient ideas of how non-harmful interaction with nature can occur in Singapore. It does not address the fundamental issue of prioritizing human development and needs over healthy and resilient ecosystems within the City in Nature vision. Nature-inclusive spaces are treated as reactions to the symptoms of a larger system that initially destroys these spaces. Hence, many conservation efforts and governmental goals of embedding more natural spaces in Singapore continue to adhere to the ideology of a hierarchical relationship between humans and nature, continuing to 'other' nature rather than challenging this tension.

This aligns with the decolonial scholar Amitav Ghosh's critique in his book *The Great Derangement* of postcolonial urban planning, and how this planning resulted in a hubristic approach to landscape planning that sanitizes and manages humanity's interactions with nature, rather than confronting nature's vulnerabilities and our complicity in exacerbating them (Ghosh 2016). This 'uncanny' (Ghosh 2016, 30) separation of human and nature, expressed and encoded in the text of the Eco-Link@BKE and other areas of the City in Nature vision, can delude communities into thinking that they are protecting the environment, which would be *beneficial*. However, in reality, these and similar texts in Singapore that uphold this City in Nature vision can conceal the ongoing damage to *Community* that humans in Singapore are complicit in through their effects on the environment and the precarity created by Singapore's development. As a result, a coded human–nature binary obscures our complicity in the subjugation of nature, preventing us from confronting the full extent of the destruction that human activity causes to the flora and fauna of Singapore. This opposes our ecosophy that includes a *Critical* approach to interrogating whether our sustainability approaches truly target the root cause of environmental destruction in the polycrisis of climate change, biodiversity loss and surpassing planetary boundaries. Because the Eco-Link@BKE perpetuates anthropocentrism and growthism by ideologically encoding a human/nature binary in its text, yet it does assist in mitigating some harms to some faunal species of Singapore, based on our ecosophy, it is an *ambivalent* discourse on Singapore's landscape.

Defining Singapore's Green Spaces

Constructing a City in Nature remains a complex task, one that is more often than not a well-intentioned process mired with unprecedented roadbumps and complications that introduce elements of ambivalence in the City's spaces when graded upon our *ecosophy*. Nevertheless, Singapore has inarguably done a stellar job in increasing the amount of greenery we see on our roads, buildings and most facets of our urban environments. Singapore's meticulous approach to urban planning has contributed to the global recognition of the city-state's efforts as a 'green city'. Travelling throughout the city, at first glance, it often feels that the green and the concrete have seamlessly integrated as one. This appears even more so in many of Singapore's popular tourist destinations that were built according to the logics of the City in Nature discourse, such as the famed Gardens By The Bay, an urban park located in Singapore's Central Business District whose conservatories attract over fifty million visitors yearly (Figure 4.3).

As mentioned above, in 2016, the Arcadis Sustainable Cities Index ranked the nation as the most sustainable city in Asia and second in the world, falling just short of Zurich, Switzerland (2016). More recently, Singapore was ranked seventh globally and first across the Asia-Pacific region under the Global Destination Sustainability Index (Global Destination Sustainability Index 2024). These, and other forms of recognition, can and should be used as case studies to highlight efforts to create green spaces in dense urban areas, steps all urban areas worldwide should take for various anthro-and-ecocentric reasons. This is why the case studies in this chapter are categorized as *ambivalent* discourses – they undoubtedly contain *beneficial* aspects of our ecosophy, where the drive to create

Figure 4.3 The Cloud Walk atop The Cloud Forest conservatory. (Photo credit: Yock.)

and cultivate green spaces is indeed one form of caring for the environment at an administrative level. However, this approach is not infallible and comes at certain costs, which complicates this discourse, introducing elements of *ambivalence*.

Two key tenets underpinning the City in Nature's developmental logic to increase green space across Singapore are the development of over 130 hectares of new parks by 2026 and the increase in the land covered by NParks management by over 50 per cent from 2020 baseline numbers (Ministry of Sustainability and Environment et al. 2021). Singapore's quest to expand its green spaces is not something new to its urban planners. It has been a constant across the discourses preceding the City in Nature project, which successfully increased green areas by 11 per cent, from 36 to 47 per cent between 1986 and 2007 (Biophilic Cities n.d.). This increase should be seen as a success, even if an *ambivalent* one. More green spaces are often better than less, although as we argue in this chapter, Singapore's approach to framing and creating such spaces is ultimately *ambivalent* as per our ecosophy.

It is vital to our ecosophy to be *Critical* – to continually question the attitudes and worldviews that influence how land is viewed. Beyond its formula, what exactly does it mean to be a green space? Can and should a green space be measured quantitatively? Does attaching a numerical value to green spaces accurately represent the relationship between human and non-human? How does the language of understanding natural space construct different *identities* for distinct facets of nature, and how does this grant *salience* accordingly?

For one, the lenses through which land is spatially represented are often subjective and dependent on the motivations of the people conducting the mapping process. Thus, to further our analysis, we need an understanding of Singapore's definition of a green space, as the notion of what it means to be a green space invites questioning. This brings about the crucial need to comprehensively understand the methodologies used in quantifying Singapore's green spaces as used by Singapore's government.

The calculation in which Singapore's green spaces increased by 11 per cent is derived from a formula defined by NParks that is grounded in green plot ratios, or the 'area-weighted average leaf area index of a site' (Tan 2009). Essentially, this definition takes the total three-dimensional volume covered by plants using the prescribed Leaf Area Index (LAI) (Urban Redevelopment Authority 2024b).

However, the fixation on quantifying green spaces can introduce a host of potential destructive issues based on our ecosophy. As hinted above, leaf area calculations ignore other important characteristics of green spaces, such as plant diversity, species interactions and habitat provision. Despite the prevalence of greenery throughout Singapore's built environment, 37 per cent of Singapore's native flora and fauna has vanished due to deforestation and urbanization in the past two centuries (Chisholm et al. 2023). Most importantly, forests are increasingly razed to make way for concrete jungles. How can we then continue to justify the 'green spaces' within cement cities as replacements for nature, simply because they encompass an arbitrarily determined amount of 'nature' as defined by 'green'=leaf area?

Additionally, obsessing over the terminology of a 'green space' fails to recognize that nature is much more than green, creating an *erasure* of vital aspects of biodiversity. This *erasure* and *framing* of green spaces as only being leaf area discounts the many other natural habitats that are not green but should still be valued, such as marine or 'blue' ecosystems, and other faunal or microscopic species, such as local fishes and the ecological richness of mudflats, intertidal areas, coral reefs and anything hidden underwater.[7] Despite their absence of 'green', they still have essential contributions to healthy ecosystems of Singapore via important pathways such as microclimate regulation, biodiversity, or soil health. Per Stibbe, the misplaced overemphasis on green plot ratios 'contains only the faintest trace of the natural world, and no hint of intrinsic value', which could end up being counterproductive and discouraging the protection of natural spaces (2021). Such *patterns of erasure* are classified as *traced erasure*, and as they continue to encode vital processes of ecosystem assessment, they engender a story that forests in Singapore only have value based on how they contribute to leaf volume and thus, define green spaces.

Furthermore, relying on canopy areas to represent green space favours larger trees, undervaluing smaller plants, shrubs and ground cover flora like mosses. Many of the large, leafy trees that line Singapore's roads are also non-native species, such as the trumpet tree (*tabebuia rosea*) from Mexico and the rain tree (*samanea saman*) from South and Central America (Lee 2020). While it is essential to avoid endorsing nativism in ecology or the idea that non-native species are undesirable, the combination of Singapore's widespread biodiversity loss and the ubiquity of alien species resulting from urban greening underscores potential shortcomings in the strategies employed to expand green spaces. On a broader scale, this quantitative preference for specific aesthetics of nature is rooted in anthropocentric developmental logics, contributing to the *voided erasure* of the intrinsic values of nature that go deeper than metrics, numbers and values.

Ultimately, the excessive focus on the numerical values of expanding green spaces raises the critical question of whether this approach fosters a substantive and meaningful relationship between Singapore's urban dwellers and the spaces they occupy. This inquiry calls for a deeper examination of the qualitative aspects of green spaces, such as their capacity to support human–nature interactions underpinned by values of radical care and reciprocity as per our shared ecosophy. Merely increasing the area of greenery without addressing the composition and functionality of these spaces poses a risk of creating an illusion of ecological progress without failing to address crucial ecological and social dynamics of the spaces, let alone critiquing underlying ideologies of anthropocentrism and growthism.

It is thus crucial to investigate the pillars supporting Singapore's construction as a 'City in Nature', as centring biophilia (National Parks Board 2021), and as a result, biophilic architecture, has been a key driver of Singapore's attempt to green its spaces, one that is encoded in Singapore's institutional policies towards transforming Singapore into a 'City in Nature'. The term 'biophilia' is rooted in two ancient Greek words, *bio* meaning 'life', and *philia* meaning 'love', literally translating into a love for life. Biophilia therefore refers to the innate human affinity for nature and natural systems as hypothesized by biologist E.O. Wilson (Wilson 2021). However, biophilia as a developmental logic that manifests in Singapore's urban spaces forms a *destructive discourse* when power structures prioritize certain aspects and types of life over others, privileging *salience* in the form of 'love' to curated, ornamental forms of nature that align with economic or aesthetic goals while disregarding ecological integrity and biodiversity. This selective *framing*, deficient in careful consideration for the qualitative, intangible needs of nature, often *erases* local ecosystems and commodifies nature, reducing it to a resource for human use rather than fostering a truly reciprocal relationship with the natural world of which we are part.

Despite the critical examination above, it is still vital to recognize that Singapore has done many things right when it comes to urban greening, where there is an effort to incorporate both the priorities of ecocentric logic and anthropocentric logic and that these successes owe their existence to the nation's overarching developmental discourses. However, the ambivalence of this discourse lies in the linguistic construction of Singapore as a 'City in Nature', where the development of Singapore's contemporary existence and visions for the future as a 'City' is *foregrounded*, *framing* urban expansion as the primary land-use priority. Regrettably, this comes at the expense of 'Nature', whose entities are relegated to the *background*, with their diminished *salience* treated as an afterthought rather than being placed on equal footing by those who determine the future of the space.

This calls to attention the language we use when referring to and building relationships with our spaces. When terminologies underscoring areas such as forests or open-space plots surround the ideas of 'vacancy' or 'underutilization', this introduces the implicit notion of their inevitable development rather than the preservation or appreciation of these vital ecological zones (Government of Singapore 2024). These and similar terms perpetuate *destructive stories-to-live-by* where the human is still centred and other organisms and ecosystems are only valued in how they can provide services to Singapore-as-City. Ultimately, such language fails to recognize the inherent ecological value that natural land possesses, *framing* land instead as more valuable when it provides services to human communities.

In a small city-state limited by land constraints, Singapore has faced inevitable negative externalities to development. This case study does not contest the value of incorporating greenery within Singapore's urban landscape and recognizes that policymakers and institutions have been forced to negotiate trade-offs between supporting a growing nation and managing its natural ecosystems. As a result, this case study instead advocates for a *reframing* of this crossroads and hopes to imagine a shift in attitudes and worldviews where development need not be a mutually exclusive phenomenon from ecological protection and conservation.

Corridors of Communities and Conservation in the City

Despite the above criticisms, the 'City in Nature' discourse has born stories that are deemed successful with regards to our ecosophy, with the Rail Corridor being exemplary of its *beneficial* aspect. Formerly the Keretapi Tanah Melayu (KTM) Railway Line, the Rail Corridor is one of Yock's favourite places to frequent in Singapore, and holds significant historical and cultural importance for Singapore. Stretching 24 kilometres from the Northern to the Southern parts of the island, it once served as a vital transportation link between Singapore and Malaysia until its closure in 2011. The last train departed from Tanjong Pagar Railway Station, which is now a National Monument, marking the end of an era and the beginning of a new lease of life for vital spaces across Singapore's landscapes. Its closure is also symbolic of the tensions in petrocultures between public transit and fossil fuel-based internal combustion engines, and how each mode of movement requires drastically different built infrastructure. This symbolism and the open-endedness of the Railway Line's future were initially met with a large wave of fervour in the form of community activism and public interest following its closure. Recognizing the corridor's potential as a green space, the Nature Society of Singapore (NSS) proposed its preservation as a continuous 'green corridor', which garnered widespread support which culminated in the creation of a digital platform titled thegreencorridor.org which acted as a platform for Singaporeans to share their lived experiences and relationships with the space and to advocate for its future (Green Future Solutions n.d.).

In response to this public enthusiasm, in March 2015, the URA launched a Request for Proposal (RFP) to develop a Concept Master Plan for the Rail Corridor. Through extensive community engagement efforts, this RFP sought innovative design solutions that would transform the corridor into a community space while integrating new housing and activity nodes. This included workshops, public exhibitions and an online portal to gather feedback and ideas from residents about the future of the Rail Corridor. The formation of the Rail Corridor Consultation Group (RCCG), comprising various stakeholders including environmentalists and local residents, was pivotal in shaping conservation strategies and ensuring that community voices were heard in planning discussions.

Eventually, this led to plans to imagine the spaces of the Rail Corridor in consideration of public feedback that heavily emphasized the conservation of the natural and cultural heritage of the space, where both are made *salient* in the corridor's redesign. One such example of public feedback called for the *reframing* of the corridor from merely a site of nature conservation to a site that centres ecological education and advocates for nature conservation. With beneficial alignment with our shared ecosophy of expanding *Community* beyond the human to include non-human nature in order to extend care in just and equitable manners, this feedback appealed for the space to facilitate providing people with the understanding of the importance of their interrelationships with Nature, allowing for a stronger affiliation and reconnection with Nature (jimmytst 2015) (Figure 4.4).

Overall, the co-creation process for the Rail Corridor involved collaborative planning between government agencies and community members to develop a shared vision for its future. This included launching initiatives such as the 'Journey of Possibilities' Ideas Competition, which invited public submissions on enhancing the corridor's use as a community space. Over 200 proposals were received, reflecting a diversity in Singaporeans' perspectives on biodiversity, heritage celebration and community ownership. The URA also engaged educational institutions to involve youth in workshops that emphasized hands-on learning about urban design.

Figure 4.4 Commuters enjoying an evening stroll along the Southern half of the conserved Rail Corridor, surrounded by lush foliage. (Photo credits: Yock.)

This approach demonstrates how developing urban greenery under the overarching logics of the City in Nature discourse can align with our ecosophy as a *beneficial discourse*. Encompassing key tenets of *Community*, this not only prioritized inclusivity and, in doing so, enriched the planning process but also ensured that the corridor would serve as a vibrant public space that meets the needs of various user groups (including non-human users). It also evidences the possibilities in Singapore as a smaller nation where transparent processes can be created and provide opportunities for active civic engagement, where these processes can be utilized to generate pathways beyond strict state-controlled ecomodernization. Critically, this approach imbued the members of the public in Singapore with the agency to *reframe* the spaces of the Corridor and decide how and to whom *salience* was privileged.

Today, this corridor has transformed from a physical barrier that divides neighbourhoods into a green space that connects communities, fostering social interaction and ecological bonds amidst Singapore's urban landscape. Notably, approximately one million residents live within 1 kilometre of the Rail Corridor, with this number encompassing public housing estates, condominiums and landed private homes, highlighting its accessibility and relevance to a large segment of the population.[8] The Rail Corridor also represents a significant victory for citizen participation and civic engagement in Singapore. Having actively involved residents in decision-making processes, making this a participatory-planning approach to shaping spaces, the Singaporean government was able to reflect the needs and aspirations of diverse communities, which foster greater personal connections to ecological spaces and promote the dissolution of human and non-human nature binaries (Choi et al. 2023). Crucially, this initiative exemplifies how collaborative governance can lead to innovative urban solutions that respect and preserve historical legacies while addressing contemporary needs. Such collaborations will only continue to be important to Singapore in its climate-changed future.

Moreover, this collaboration between top-down and bottom-up approaches encouraged a rethinking of heritage and natural spaces; rather than being seen solely as areas ripe for modernization, these spaces can be appreciated for their traditions and histories. The Rail Corridor as text proves that historical spaces can be preserved in their current form while still providing value to both human and non-human communities, benefiting all in diverse ways. Ultimately, the Rail Corridor highlights a means to resist progress as is ideologically defined by modern worldviews by instead achieving progress that is ideologically defined by those on the ground. As Singapore continues to evolve, the Rail Corridor is a testament to successful citizen efforts in shaping public spaces that reflect collective aspirations while honouring ecological integrity.

The Hypercuration of a Green Aesthetic

An overarching, wider discourse with significant implications on the way land use is *framed* within Singapore is the origins of the City in Nature metaphor as a top-down imposed vision hypercurated by the state and governing bodies that reinforce the image of a green Singapore. While the state's investment in incorporating natural spaces and considering biodiversity in land-use planning can be interpreted as a shift towards ecocentrism, it conceals other intentions of strategically presenting Singapore as a frontier of new eco-cities and eco-progressiveness (Pow 2018). This upholds a specific 'green aesthetic' of Singapore as an urban utopia and that brings with it political leverage for the country on the global stage. In this *framing*, Singapore is an ecomodern exemplar of the future from whom other countries, especially in Southeast Asia, but also from around the world, can learn. The hypercurated statist City in Nature story elides and *erases* the biotic and abiotic features of Singapore, without which the humans of Singapore cannot exist. It thus reinforces the ideology of anthropocentrism, which goes against our ecosophy's pillar of *Care*, because it upholds a relationship of human domination and exploitation of the non-human.

While such a green aesthetic might not necessarily be *destructive* in and of itself, it presents a paternalistic pattern of conservation that echoes colonial relations to the land and the people (Survival International n.d.). This fails to heal the fundamental relationship between the people of Singapore and the rest of nature of Singapore and presents instead a story of nature as another commodity to exploit for the politics of aesthetics (Pow 2018). In fact, it promotes nature as an entity to be managed and organized, building on a long history of colonial ideologies that have shaped the texts of Singapore, including the world-famous Botanic Gardens.

Singapore Botanic Gardens: Exploring the Colonial Roots of the 'Garden City'

Anjali Venketram

Despite the sweltering heat and heavy humidity of a Sunday afternoon, the Singapore Botanic Gardens hum with life. Housing over 10,000 species of flora and fauna, the gardens span eighty-two hectares of prime land in the heart of Singapore. People gather throughout the gardens – on patches of grass, along pathways and by the lakes – sunbathing, jogging, playing and more. If the weather is good, you will struggle to find a place to set down your picnic mat on the grass. The atmosphere is vibrant as people come together for a day out to reconnect with nature.

Known as the 'Garden of Perpetual Summer' (Tinsley 2009), this space is one of both leisure and luxury for Singaporeans and non-residents as well, whether tourists or foreign workers. Young children strolling through the gardens stop to explore plants, guided by carefully placed signage that identifies and describes each species. As I write this analysis, the Gardens are hosting a fascinating exhibition on the ethnobotany and diversity of figs, offering insights far beyond what one might expect to know about these fruits. Manicured grass fields and curated spice gardens fill every corner. The only real uncurated elements around these gardens are otters, lizards and the growing chicken population – and perhaps the children who can often be seen chasing the latter (Figure 4.5).

Figure 4.5 Image of the entrance of the Singapore Botanic Gardens. (Photo credit: Todd.)

A portion of the Gardens preserves one of the island's few remaining patches of primary rainforest, as well. Through this primary forest runs a man-made walkway to allow a smooth stroll through the forest – reiterating the very 'well-curated' image of the Gardens, a sort of microcosm of the Garden City and, now, City in Nature itself. The presence of this path encourages more people to take a 'stroll through an ancient rainforest in the heart of the city' which the National Parks Board of Singapore has labelled as an 'accessible living labouratory and museum of natural history'.[9] This language raises questions about the relationship between Singaporeans and 'nature'. Has the forest been reduced to a labouratory and museum? Is this indicative of the fact that we are unable to see the forest for its innate value because deforestation was a major characteristic of Singapore's history and will continue to be a phenomenon Singapore grapples with in this century of planned development, too? Is this the reason why we over-romanticize and curate our access to (limited) nature in our parks, including at the Botanic Gardens? Is there an explanation for why we view nature as something that needs to be protected, set aside and romanticized?

Therein lies the *ambivalent* paradox of the Botanic Gardens: while it stands as a maintained and meticulously designed space, it also raises questions about the balance between preserving the innate experience of nature and the heavily curated and managed experience being offered to Singaporeans. One of the main drivers of the field of ecolinguistics is the need to think critically about why our relationship with nature is the way it is. So far in the book, we have discussed the *destructive* and now *ambivalent stories-we-live-by* in Singapore, and it can be concluded that stories are never just black and white – there is potential for a space to hold multiple interpretations. Discourses are standardized ways that particular groups in society use language, images and other forms of representation (Stibbe 2021, 22). The narratives and language that frame and shape the Singapore Botanic Gardens are equally complex and ambivalent, as is its history from being first a colonial pet project and now a prideful UNESCO site, such that these gardens have meant many different things to multiple generations of Singaporeans and foreigners alike.

Through this analysis, I wish for the readers to walk away with a deeper understanding of the influence that colonization had on the Singapore we know today. As mentioned in Chapter 1, Singapore was stripped of ninety per cent of its natural rainforest for spice plantations long before the country was even established in 1965 or before the government even took control of the greening of the city. The British set the original conditions for the establishment of the city and that is a fact we cannot ignore when we criticize the current world we live within, as the limits and possibilities of a truly sustainable Singapore are unquestionably shaped by this British colonial past. Part of that history includes the creation of the Botanic Gardens. Given this, I choose to speak of the colonial discourses in this part of the case study because the Botanic Gardens are a present-day remnant of England's colonial and imperial greed and exploitative regime, where this has shaped the Gardens as an *ambivalent* text. In order to understand and appreciate the Gardens today, we have to acknowledge the backdrop against which they were established. Within this regime, the Gardens became a place made for 'research' where the lines of scientific, economic and geopolitical power became completely blurred, and to this day, despite UNESCO and tourist book accolades, we are unable to distinguish these lines.

The History of The Singapore Botanic Gardens

The Singapore Botanic Gardens boasts a long history and is the only tropical garden in the world to be labelled as a UNESCO World Heritage Site. Plans for the Botanic Gardens were materialized as early as 1822 – only a few years after Sir Stamford Raffles' 'discovery' of Singapore when he established the island as a British trading post in 1819. His arrival marked the foundation of colonial Singapore when he intruded into the family politics of the Sultanate of Johor. He offered the exiled Tengku Hussein recognition as the rightful Sultan of Johor and a yearly payment in exchange for the rights to establish a trading post in Singapore. Their agreement was ratified with the Treaty of Singapore, and this established the foundation for what we call 'modern' Singapore (Antiques of the Orient 1993).

Raffles was displeased with the initial 'management' of Singapore under William Farquhar, so he set up The Raffles Town Plan to execute his vision for the urban planning of Singapore. Singapore, the city now known for its urban planning that extends years into the future, found its first plan created by its British colonizers – that too by a man who had a very specific and economically-driven vision. This plan was fraught with historical implications for Singapore – and laid out some of the doctrines that affect societal and also human–nature relations in Singapore today. For example, Raffles allotted specific patches of land to each segment of society ensuring their separation as decided by their race (Europeans, Chinese, Malay and Indian). In his plan, Raffles also emphasized the importance of having greenery – to give off an impression of prosperity. From its establishment, the Botanic Gardens also and strategically had economic motives attached to it – not to educate, not to embrace greenery for its innate value – but to add a performative element to the city, including the performance of British colonial control over the fauna (and peoples) of Southeast Asia. This emphasis on greening was ironic, coming from the very regime that was clearing vast swathes of rainforest to establish cash crops.

Raffles was drawing from his experiences with other colonial towns where he saw economic success – like Calcutta in India – and transplanted the same ideas into Singapore. His ideas established early on that there was a clear economic intention behind greening a city. For him, the green space represented an economic opportunity – making the city more attractive to trade and investment. Lee Kuan Yew held the belief that green spaces encouraged foreign investment in a city when it came to his long-term plans for Singapore (National Parks Board, n.d.). There is no doubt that many colonial practices had a long-term effect on the governance of Singapore and that the ideologies of anthropocentrism and growthism have been central to the design and management of Singapore as a first colonial outpost and now as an independent country.

It has also long been known that, unlike some other cities that scorned their British rulers for the unjust and immoral rule in other colonies (Bombay became Mumbai), Singapore celebrates its colonial rulers. It was described in succinct terms by Singaporean poet Gwee Li Sui who said 'While around the world, colonial statues are coming down, here, they are going up' (Yong 2024). In 1969, mere years after Singapore established its independence, a statue of Raffles was placed in the heart of the city – an ode to the 'founder' and trailblazer to establish Singapore as a location of great economic significance. To this day, there are three statues of him in Singapore. In an address to Singaporeans in 2019, then-Prime Minister Lee Hsien Loong acknowledged that without Raffles' plans for Singapore, the nation would not be where it is today, claiming that '1819 marked the beginning of a modern, outward-looking and multicultural Singapore' (Chia 2019). This speech was made during Singapore's bicentennial celebrations in 2019 – an acknowledgement of the establishment of modern Singapore in 1819.

The Colonial Implications of the Botanic Gardens

The inspiration Singapore draws from its colonial predecessors is evident in the city's planning, raising a 200-year-old question: where would Singapore be without colonialism? While this case study does not aim to answer that question, this context of colonialism is essential for understanding the relationship between present-day Singapore and its colonial past under British rule. This relationship is also essential for any ecolinguistical analysis of the Botanic

Gardens, where according to UNESCO 'The Singapore Botanic Gardens is situated at the heart of the city of Singapore and demonstrates the evolution of a British tropical colonial botanic garden from a "Pleasure Garden" in the English Landscape Style to a colonial Economic Garden with facilities for horticultural and botanical research, to a modern and world-class botanic garden, scientific institution and place of conservation, recreation, and education' (UNESCO n.d.). This relationship and its evolution are *ambivalent*, complex, and not easily condemned, given the perceived benefits of the policies implemented during colonial rule. While we discuss the 'positive' influence of British colonial greening policies, we must also acknowledge the illiberal, damaging and violent traditions Singapore may have inherited from colonial rule. While Singapore celebrates the positive aspects of its colonial past, many other implications have been *erased*, and such *erasure* is contrary to the ecosophy that motivates this book. *Erasure* can refer to 'a story in people's minds that an area of life is unimportant or unworthy of consideration' (Stibbe 2021, 146). Histories all too often *erased* are dispossession of native groups from their lands; the fact that Singapore thrived because of the opium trade and slavery; and the fact that British governors used harsh oppressive tactics to gain and maintain control over the entire territory (Han 2018). This says nothing for the native flora and fauna also *erased* from their own ecologies of alive places that today is the location of the Botanic Gardens.

Because European colonization is at its core a process of violent dispossession in service of wealth extraction, the establishment of the Botanic Gardens was, in large part, to act as a test-bed for the cultivation of 'revenue-earning' crops and ornamental plants for Singapore and Britain's colonial landholdings throughout Southeast Asia. However, original interest in the gardens died out until the Gardens' re-establishment in 1859 – and by then, it was reported that more than 45,000 acres of land in Singapore had been ironically abandoned after being cleared for plantations. This is an important fact to note – once a primary forest is lost, it can never be recovered fully. The landscape of Singapore was characterized by indiscriminate deforestation by the late nineteenth century, with much of this driven by efforts to cultivate gambier, pepper and rubber, with such deforestation leading to the extinction of megafauna like the Asian tiger (driven extinct in the 1930s). Ironically, there was a turn in the 1870s after a government field survey yielded disappointing results because of a lack of 'legislation for forest protection' (Goh 2018). It was then decreed that many pieces of forested land be given 'protected status' including spots in Bukit Timah and Mandai that remain today.

The creation of 'nature' as something separated from humans to be revered and protected is a highly anthropocentric narrative and goes directly against the ecosophy of this book. It reeks of irony that the ones that created these protections were also the ones fully exploiting the land – a sentiment that defines Singapore's dominant view of forests to this day: they are 'resources' that are meant to be exploited and Singapore should protect them because they may deplete (a lesson from our colonizers). As Stibbe posits when speaking of ecosystems, 'when living beings become 'resources' the entailment is that not using them is a waste and that they should be exploited rather than left to get on with their own lives' (Stibbe 2021, 157). Thus the need to 'conserve' land – to this day Singapore applies this conservation narrative and thus the ideology of human/nature separation to the last remaining patch of primary forest within the Botanic Gardens. Conservation is a colonial construct – it is an inherited ideology that via governmental policy seeks to control the ecosystems of Singapore and to assert human control on the use of the remaining patches of forest. NParks policy can be ideologically traced back to the very same narratives of economic exploitation, human control and aesthetic appreciation embedded in Singapore by its British colonizers. Within this lineage, the inherently more 'modern' practice of conservation was justified through 'scientific evidence' that decreed the proper management of these forests. While today we view the conservation of forests in Singapore as a requirement to 'constantly balance demands and trade-offs across a wide variety of needs, including housing, green spaces, infrastructure, community facilities, workplaces, amongst others' (Ong 2021), conservation has in the past been economical as a systemic form of forest management that ensured profitable extraction for export European markets. This entire colonial history is too often erased from the City as Nature narrative, and also from all of the global accolades that posit Singapore as an ecomodern and green city of the future.

Rather, to this day, we see the Botanic Gardens as a space where we can 'extract' knowledge; we see the primary forest as a 'test-bedding' site. We force ourselves to see the depletion of these ecosystems as a choice – should we leave a patch of forest or ensure housing for our citizens? The question is troublesome to most who would go for the conservation of slivers of forest (such as the one in Botanic Gardens) to justify that we did leave some forest and ensured somehow the protection of the benefit of the 'resource' that forests are to us. Slavoj Zizek called ecology the 'opium' of the masses which we use to soften the reality of life under the crisis of capitalism we are going through (Zizek 2008). NParks prides itself on creating the 'best living environment through excellence in nature conservation' (National Parks Board, n.d.). There is no doubt that it has achieved this *excellence* – but, we must question the pat on the back we give ourselves for a practice that is inherently construed with extractive ideals; are such pats a pacifying environmentalist capitulation to human supremacy?

When the park regained exposure in the late 1870s, Nathaniel Cantley developed the 'Economic Garden' in 1879 within the gardens to conserve and research plants with economic potential (Lim 2012). Rubber seedlings bought in from England kicked off rubber cultivation at the gardens and a spark of rubber plantations across Malaya. The Botanic Gardens acted as a tool for the propagation of the imperial strategy because the research fuelled the deforestation of peninsular Malaya. There was a boost in research activities like orchid hybridization that led to the creation of Singapore's national flower the orchid hybrid Vanda Miss Joaquim. Now, the Singapore Botanic Gardens is one of the world's top centres for orchid growing and boasts a large Orchid Garden with many engineered varieties of orchids. In present-day Singapore, these orchids are used for 'orchid diplomacy' whereby the scientific prowess of the gardens boosts Singapore's bilateral diplomacy – often naming new varieties of orchids after national leaders when they visit (Tan n.d.). In this way, there is an implication that scientific discoveries and the manipulation of nature can provide tangible economic and political benefits to the nation – which can further the harmful narrative that nature can be manipulated and exploited for anthropocentric benefits.

Although there are some narratives of the Gardens that are wrought in structures of power, the Gardens are indeed aesthetically beautiful and scientifically important – there is no denying the positive effect the space has on Singaporeans. Without a space to meet and have picnics, to let dogs run free and to see every kind of flora imaginable, Singapore would be at a loss. It cannot be denied that NParks puts a lot of effort into the education of the public about the environment where besides making sure there are signs that explain the significance of flora and fauna at the Gardens, NParks also hosts guided tours for adults and children (National Parks Board n.d.).

However, when judged by our ecosophy then the human use of the natural world at the Gardens is largely based on past and present ideologies of separation. And despite the positive educational aspects of the Gardens we only require this level of education because as residents of Singapore, we lack other sources of interaction with nature at scale (Ngo, Hosaka and Numata 2019). It becomes hard for us to see and understand ecosystems outside of the highly urban and managed context we live in in Singapore – where lush forests don't extend for miles on every end. Many aspects of the success of the Gardens can be attributed to the British Empire and the efforts they took to establish the gardens. However, current mainstream discourses about the Singapore Botanic Gardens are overwhelmingly positive – as tourism websites celebrate, 'from school children to avid botanists, there is something for everyone to learn and explore' (Thursd 2024). The Botanic Gardens is also going to have the largest plant database in Southeast Asia by 2027 – becoming a crucial site for researchers and the scientific community (Yu 2024). All these positive narratives help with framing the Botanic Gardens as a *beneficial* space. Clearly, and for sure on the surface, given such framing, the Gardens can be considered a *beneficial* space. This is so even to someone with the most intentionally inclusive ecosophy because the *destructive* aspects of the Gardens can be hidden under a facade of what conventional environmentalism looks like for it to be considered successful.

The core tenets of our ecosophy are *Community*, *Care* and *Critical* thought – and I believe it would be contrary to our ecosophy to label the space of the Botanic Gardens as a merely *beneficial* or *destructive* discourse. The value of this famous and important 'text' of Singapore is deeply nuanced and embedded within the context of the country's colonial past and its present urban history. There are also genuine and hopeful narratives attached to the gardens, especially from younger generations as they grow up needing to co-learn with the spaces they grow up in. But what I, and more of us as we read through this book, need to recognize is the fraught colonial history it represents. Built for the economic prowess of the British Empire, we cannot separate it from the benefits it brought to our colonizers at the cost of our environment – and in many contexts, the cost of livelihoods and lives themselves. It set the scene for the economic success we strive towards as a nation and we need to be cognizant of the narratives and thus ideologies that envelope our logics of sustainability and that have the potential to taint our sustainable futures as well. At the end of the day, we are left to weigh the pros and cons of what space means to us according to our ecosophies. As Stibbe posits, 'we will need new stories to survive by' (Stibbe 2021, 183) and we need to tell a new story about the Botanic Gardens that is celebratory and positive while also being cognizant of the implications of its colonial past and managed present.

Conclusion

Tan Yock Han

The vision of this chapter, and more broadly, this book, is the development of a society that holistically accounts for the needs of both human and non-human nature, avoiding the binary or pitting of the two against one another. This vision is illustrated through the case studies of the Singapore Botanic Gardens and the 'City in Nature', two distinct yet interconnected discourses on urban–nature relationships that elucidate factors such as space, time and positionality as variables that shape the ambivalent relationship between Singapore's discourses and our ecosophy.

These case studies highlight the intricate and sometimes contradictory dynamics of urban development in Singapore. They highlight how *ambivalence*, marked by the coexistence of *beneficial* and *destructive* elements, can both inhibit and inspire the development of a sustainable urban future. This highlights how it is paramount to develop a more conscious and nuanced approach to understanding our spaces – especially one that creates an open-ended future that encourages community engagement and engagement with the more-than-human in order to nurture a sustainable and reciprocal relationship between Singapore's urban population and its natural environments. However, it is precisely this *ambivalence* that makes this chapter unique and important. In showcasing the imperfections, this coexistence of the *beneficial* and the *destructive* together, this critical approach to Singapore's discourses is vital towards realizing the core of our ecosophy **'Majulah…?'**, of making space for more *beneficial* discourses to grow to create a better future for all inhabitants of Singapore's spaces.

Before ending this chapter, we leave you with some reflective questions we feel are imperative to this chapter's themes:

- What spaces or ecological ideologies do you hold in high regard, even as their imperfections (or *ambivalence* to your personal ecosophy) have become apparent over time? How has this awareness influenced or redefined your relationship with them?
- What do these case studies teach us about discerning *ambivalence* from *beneficial* discourses? A key theme in our chapter is problematizing things that might initially seem *beneficial*, but upon looking at the ideologies they reify, we can then identify what discourses can actually build towards a more hopeful climatic and sustainable future.

Notes

1 For the sake of brevity, different definitions of gardens are not explored here. Instead, we want to focus on the overall connotation of a garden in the

contemporary context as a space curated by humans for human benefit. Gardens can be defined as bridging spaces by humans of non-human Others (Alexander 2013) but where this relationship is still anthropocentric by design.

2 Taken directly from NParks' publicly available website.

3 All numerical goals are qualified to be additional to existing spaces.

4 Wu 2023 stated that, 'The future loss of secondary forests is about 1.2 times larger than the entire area of all parks and nature reserves in Singapore'. However, due to having no clear reference as to how the park and nature reserves area was found, the comparison can instead be drawn to the total area of NParks land – which consists of nature reserves, parks and park connectors.

5 For more on Singapore's cleanliness discourse, refer to Centre for Liveable Cities 'Cleaning a Nation: Cultivating a Healthy Living Environment' (2016) and BBC's 'Singapore's endless pursuit of cleanliness' (2021).

6 This will be expanded on in the following case study on the Botanic Gardens and Singapore's colonial histories of nature.

7 Readers can refer to the case study Chek Jawa for more on histories of wetlands in Singapore.

8 See Chapter 5's case study on Singapore's Park Connector Network for a deeper analysis on the significance of park connectors throughout Singapore, of which the Rail Corridor serves as a key component of this network.

9 For a brief history of the Gardens as curated and shared by NParks, see https://www.nparks.gov.sg/sbg/about/our-history. Accessed 28 November 2024.

References

Alexander, Vera. 2013. 'Environmental Otherness: Nature on Human Terms in the Garden'. *Otherness: Essays and Studies* 4 (1): 28.

Antiques of the Orient. 1993. *Sir Thomas Stamford Raffles: Book of Days*. Singapore: Antiques of the Orient.

ARCADIS. 2016. 'Sustainable Cities Index 2016'. https://www.seoulsolution.kr/sites/default/files/gettoknowus/%5BArccadis_Report%5D%20Sustainable%20Cities%20Index%202016.pdf.

Biophilic Cities. n.d. 'Singapore'. https://www.biophiliccities.org/singapore.

Chia, Anthony. 2019. 'PMO | PM Lee Hsien Loong at the Launch of the Singapore Bicentennial'. Text. Prime Minister's Office Singapore. 31 January 2019. https://www.pmo.gov.sg/Newsroom/PM-Lee-Hsien-Loong-at-the-launch-of-the-Singapore-Bicentennial-Jan-2019.

Chisholm, Ryan A., et al. 2023. 'Two Centuries of Biodiversity Discovery and Loss in Singapore'. *Proceedings of the National Academy of Sciences* 120 (51): e2309034120. https://doi.org/10.1073/pnas.2309034120.

Choi, Jaz Hee-jeong, Kit Braybrooke, and Laura Forlano. 2023. 'Care-Full Co-Curation: Critical Urban Placemaking for More-than-Human Futures'. *City* 27 (1–2): 15–38. https://doi.org/10.1080/13604813.2022.2149945.

Ghosh, Amitav. 2016. *The Great Derangement: Climate Change and the Unthinkable*. University of Chicago Press. https://doi.org/10.7208/chicago/9780226323176.001.0001.

Goh, Lee Kim. 2018. 'When Tigers Used to Roam: Nature and Environment in Singapore'. *biblioasia.nlb.gov.sg*. https://biblioasia.nlb.gov.sg/vol-13/issue-4/jan-mar-2018/whentigersusedtoroam/.

Government of Singapore. 2024. 'Buying Vacant Land or Development Sites'. *Inland Revenue Authority of Singapore*. 13 December 2024. https://www.iras.gov.sg/taxes/property-tax/property-buyers/buying-vacant-land-or-development-sites.

Green Future Solutions. n.d. 'The Green Corridor'. Accessed 5 December 2025. http://www.thegreencorridor.org/

Han, Kirsten. 2018. 'Colonialism a Cause to Celebrate in Singapore'. *Asia Times*, 21 January 2018. http://asiatimes.com/2018/01/colonialism-cause-celebrate-singapore/.

Hicks, Robin. 2023. 'For Future Land Use, Singapore Could Deforest Area Larger than Its Nature Reserves and Parks Combined: Study'. *Eco-Business*, 4 July 2023. https://www.eco-business.com/news/for-future-land-use-singapore-could-deforest-area-larger-than-its-nature-reserves-and-parks-combined-study/.

jimmytst. 2015. 'My Feedback on the Concept Master Plan to Develop the Green Corridor in Singapore'. *Nature and Us* (blog), 11 November 2015. https://natureandus.wordpress.com/2015/11/11/my-feedback-on-the-concept-master-plan-to-develop-the-green-corridor-in-singapore/.

Lee, Melissa. 2020. 'Native or Not?' *NParks Buzz* (blog), December 2020. https://www.nparks.gov.sg/nparksbuzz/dec-issue-2020/main-feature/native-or-not#:~:text=Although%20the%20Rain%20Tree%20(Samanea,shaped%20and%20wide%2Dspreading%20crown.

Lim, Tin Seng. 2012. 'From Botanic Gardens to Gardens by the Bay: Singapore's Experience in Becoming a Garden City'. BiblioAsia. https://biblioasia.nlb.gov.sg/vol-8/issue-2/jul-2012/singapore-botanic-gardens-city/.

Ministry of Sustainability and Environment. 2024. 'City in Nature'. *Singapore Green Plan 2030*, 3 December 2024.

Ministry of Sustainability and Environment, Ministry of Education, Ministry of National Development, Ministry of Transport, and Ministry of Trade and Industry. 2021. 'Singapore Green Plan 2030'. Singapore Green Plan 2030. Accessed February 2021. https://www.greenplan.gov.sg/overview/.

Montefrio, Marvin Joseph F., Xin Run Lee, and Elwin Lim. 2021. 'Aesthetic Politics and Community Gardens in Singapore'. *Urban Geography* 42 (10): 1459–79. https://doi.org/10.1080/02723638.2020.1788304.

National Parks Board. 2021. 'Annex B Fact Sheet: Transforming Singapore into a City in Nature'. National Parks Board.

National Parks Board. 2023a. 'Eco-Link@BKE'. National Parks. Accessed 26 September 2023. https://beta.nparks.gov.sg/visit/parks/bukit-timah-nature-reserve/special-features/eco-link-bke.

National Parks Board. 2023b. 'Singapore, Our City in Nature'. NParks: City in Nature. Accessed January 2023. https://www.nparks.gov.sg/about-us/city-in-nature.

Ngo, Kang Min, Tetsuro Hosaka, and Shinya Numata. 2019. 'The Influence of Childhood Nature Experience on Attitudes and Tolerance towards Problem-Causing Animals in Singapore'. *Urban Forestry and Urban Greening* 41: 150–7. https://doi.org/10.1016/j.ufug.2019.04.003.

Ong, Justin. 2021. 'S'pore Must Balance Stewardship of Green Spaces with Land Constraints: Desmond Lee'. *The Straits Times*, 1 February 2021. https://www.straitstimes.com/singapore/politics/spore-must-balance-stewardship-of-green-spaces-with-land-constraints-desmond-lee.

Plumwood, Val. 2008. 'Shadow Places and the Politics of Dwelling'. *Australian Humanities Review* 44. https://australianhumanitiesreview.org/2008/03/01/shadow-places-and-the-politics-of-dwelling/.

Pow, C. P. 2018. 'Building a Harmonious Society through Greening: Ecological Civilization and Aesthetic Governmentality in China'. *Annals of the American Association of Geographers* 108 (3): 864–83. https://doi.org/10.1080/24694452.2017.1373626.

Stibbe, Arran. 2015. *Ecolinguistics: Language, Ecology and the Stories We Live By*. London: Routledge.

Stibbe, Arran. 2021. *Ecolinguistics: Language, Ecology and the Stories We Live By*. Second edition. London: Routledge, Taylor and Francis.

Survival International. n.d. 'Indigenous People Are the Best Conservationists'. *Survival International*. https://survivalinternational.org/campaigns/conservation.

Tan, Puay Yok. 2009. *Leaf Area Index of Tropical Plants: A Guidebook on Its Use in the Calculation of Green Plot Ratio*. 2nd ed. Centre of Urban Greenery and Ecology.

Tan, Sec. 2018. 'Singapore's Environmental Policy: Charting Progression in the Island-State'. In *Routledge Handbook of Sustainable Development in Asia, 1st Edition*, 91–103. London: Routledge, Taylor and Francis.

Thursd. 2024. 'Exploring the Singapore Botanic Gardens'. *Thursd*, 31 July 2024. https://thursd.com/articles/singapore-botanic-gardens.

Tinsley, Bonnie. 2009. *Gardens of Perpetual Summer: The Singapore Botanic Gardens*. Singapore: iSingapore Botanic Gardens.

UNESCO World Heritage Centre. 2023. 'Singapore Botanic Gardens'. Accessed 26 November 2024. https://whc.unesco.org/en/list/1483/

Urban Redevelopment Authority. 2024b. 'Greenery'. *Urban Development Authority*, 20 December 2024. https://www.ura.gov.sg/Corporate/Guidelines/Development-Control/Non-Residential/SR/Greenery#:~:text=The%20Total%20Leaf%20Area%20shall,or%20scientific%20names%20of%20plants.

Waves of Transformation 2024 GDS-Index Report. Vol. 2024 GDS-Index Report. Global Destination Sustainability Index Report, n.d.

Wilson, Edward O. 2021. *Biophilia*. Cambridge, MA: Harvard University Press.

Wu, Yitong. 2023. 'Study towards Integrating Secondary Forests into Future Land Use Development in Singapore for Biodiversity Objectives'. *Sustainability* 15 (4): 2916. https://doi.org/10.3390/su15042916.

Yong, Clement. 2024. '"Colonialism Is Not Neutral": Third Public Statue of Sir Stamford Raffles Ignites Online Debate.' *The Straits Times*, June 2024. https://www.straitstimes.com/life/arts/colonialism-is-not-neutral-third-public-statue-of-sir-stamford-raffles-ignites-online-debate.

Yu, Ariel. 2024. 'World of Plants a Click Away: S'pore to Boast S-E Asia's Largest Digital Plant Database around 2027'. *The Straits Times*, 6 August 2024. https://www.straitstimes.com/singapore/world-of-plants-a-click-away-s-pore-to-launch-s-e-asia-s-largest-digital-plant-database-around-2027.

Zizek, Slavov. 2008. 'Nature and its Discontents.' SubStance 37 (3): 37–72.

5

Beneficial Discourses of Singapore

Beverley Choo Jia Ying, Hannah Ng Shueh Yi and Kristina Gweneth Ponce Simundo

Introduction

Hannah Ng Shueh Yi

The core of the ecosophy posited in Chapter 2 is ***Majulah…?***, inherent in which is a question for Singapore's future. *Beneficial* discourses provide examples of possible plural sustainable futures. They offer an avenue to search for stories with ideologies commensurate to the climate-changed era, pointing towards ways to build upon these existing stories, and most importantly, provide foundations for those futures which better align with deeply sustainable and environmentally oriented lives for humans and non-humans alike. Stibbe used his book to highlight how linguistic features are used to create discourses, *beneficial, ambivalent* and *destructive* by social, cultural, corporate and individual forces. The global (and specifically, Singapore's) consumerist present, built upon the ideologies of anthropocentrism and growthism that inform its ecomodern ecomanagerialism unfortunately privileges *destructive* discourses too numerous to list in this book. This is why *beneficial* discourses play an important role in offering viable alternatives.

The core of ecolinguistics and its method of evaluating discourses and narratives for their alignment with an ecosophy is inspired by the normative goal of using these discourses to create a better future. Soft examples of using *beneficial* discourses to reinforce positive environmental outcomes can be seen in independent movements such as the Good Stories Movement and other climate-positive/net-positive affirming reporting sites. The Good Stories Movement was an open-sourced, expert-validated site that recognized positive climate impact around the world in 2022. Founded by one of Asia's leading voices in environmental law, Antonio A. Oposa, Jr., most well-known for fighting a decade-long legal battle of Oposa v. Factoran which helped to preserve 800,000 ha of ancient forest in the Philippines, his legacy – including the Good Stories Movement[1] – established legal protection of future generations of people (legal intergenerational protections which allow for lawsuits infringing on their rights to life) and a ban of logging of virgin forests in the Philippines. Oposa's dedication not only to legal protection but the overturning of negative discourses closely echoes Stibbe's definition of ecolinguistics which is 'critiquing forms of language that contribute to ecological destruction, and aiding in the search of for new forms of language that inspire people to protect the natural world' (Stibbe 2015, 1).

Our previous case study chapters examined various *destructive* and *ambivalent* discourses of Singapore. Those discourses are examples of those that when measured against our ecosophy must do better in creating ideologies and worldviews that move Singapore towards a feminist degrowth and strongly sustainable future. Deconstructing those discourses and highlighting the areas where those Singaporean (city) stories have reflected colonial, imperial and plantation logics and stories of dispossession of human and non-human natures per Stibbe must be challenged.

Despite having the most number of case studies of the chapters, it was difficult for us to choose examples of *beneficial* discourses in Singapore for several reasons. First, as per human expectation, nothing is ever quite good enough. However, as Stibbe acknowledged, *beneficial* discourses are the stories that need to be promoted and that are sadly not part of mainstream narratives – if they were, the world would look very different and have a vastly different society, one that is inclusive, circular and non-consumer oriented (as per our ecosophy). Since they are not mainstream, it posed a particular difficulty in identifying and presenting below compelling *beneficial* discourses. Additionally, many examples fit our ecosophy to a certain extent, but fell short of fully fulfilling all our hopes or expectations of what a deeply sustainable space should look like. There are many more examples of *ambivalent* discourses in Singapore than *beneficial* ones; given petroculture, the urban metabolism and contradictions of global capitalism, this is not surprising, both for Singapore and any other urban area the world over. On top of that, as residents of Singapore and Singaporeans, we are keenly aware of the shortcomings of our society. This is both inculcated by a strong sense of critical thinking, but also of despair that, despite the best efforts from the whole of our society and the praise of many others, we still fall short in the face of the climate crisis.

In this chapter, we endeavour to find our most treasured Singapore spaces, spaces that have been our beacons of hope, embodying the future we hope to see. We present here three of what we feel contain the foundations of the strongly sustainable future we would like to see. They are case studies of community action, of day-to-day Singaporean interactions with nature and of ecological conservation – each of which presents a different face of Singapore's *story-we-live-by*.

An ecolinguistical analysis of spaces, especially by a group with differing opinions, histories and experiences of the space and of Singapore, has also led to what could be contrasting discourses brought from different aspects of the same space. The Park Connector Network, for example, is a key tenet of the City in Nature vision – however, in alternate analytical lenses, we see in it the possibilities and workings of the future we hope for. As Stibbe points out, the criterion for *beneficial discourses* is not dominance, but instead 'new discourses that convey ideologies which can actively encourage people to protect the systems that support life… The purpose for analysing beneficial discourses is to promote them as useful alternative ways of telling stories about the world and help them to become widespread, even if they are currently relatively unknown' (Stibbe 2015, 30).

We want to acknowledge that Singapore can, and hopefully will be, brimming with *beneficial* case studies for deeply sustainable exemplars of **Majulah…?** beyond our case studies. As Singapore continues to grow and form in the pursuit of a climate-resilient City in Nature, future ecolinguistical analysis will hopefully be spoilt for choices of *beneficial* discourses.

Singapore's Park Connector Network (PCN): Everyday Stories of Hope

Kristina Gweneth Ponce Simundo

The etymology of the word 'park' stems from notions of enclosure, or areas fenced off (Oxford English Dictionary 2024). Contradictingly, 'connector' stems from connect, which is to bind together or to establish relationships between (Oxford English Dictionary 2023). Putting these two words together, Singapore named the Park Connector Network (PCN), a space that gives rise to a concoction of various discourses that shape the *stories-we-live-by*.

The dual nature of its etymology reflects the unique discursive dynamics of the PCN. The PCN is an island-wide network of park connectors that run through the entire city, interlinking various green spaces (set to be fully developed by 2030). It was originally developed in 1992 to meet the growing need for alternate recreational activities for residents in Singapore (NParks 2024). The PCN consists of four main networks of park connectors: (1) Eastern Corridor, (2) North-South Corridors, (3) Bukit Timah-Rochor Green Corridor and (4) Round Island Route. These park connectors are described as 'green corridors' that consist of paths for walking and cycling and are lined with trees and shrubs (NParks 2024). They are important loci of human and non-human interactions in the urban day-to-day of Singapore residents – and thus, the PCN plays a significant role in Singapore's sustainability story (Figure 5.1).

When looking to my own experiences of the city for examples that could express our ecosophy's focus on *Care*, *Community* and *Critical* values, I found myself thinking back to the mundanity of these interactions within the park connectors, and how they can be sites of burgeoning hope for a deeper kinship[2] between urban residents and the rest of the ecosystem. The space I feel most kinship with sits in the eastside of the country – the Pasir Ris Park Connector, which I walk through every day to access the train station that is a part of my commute to the central area of Singapore. In the process of writing this book, I began to sit with this innocuous part of my commute and found myself reflecting on the way my body moved through space and interacted with others, both the human and non-human, within the park connector. I began to see a mirage of the world our ecosophy aspires to shining through (Figure 5.2).

Hence, I delve into this case study as a way of writing to uncover the glimpses of potential for a changed relationship between urban residents and the rest of the ecosystem – one that is just and sustainable. In the section *Park*, I elabourate on how the PCN promotes a discourse of an alternative, anti-capitalist relationship to space, time and nature. Subsequently, in *Connector*, I analyse the signs in the PCN that bring to *salience* (Stibbe 2015) the interconnectivity of residents within the larger ecosystem as a whole.

It is crucial to note here that these *beneficial* discourses remain merely as glimpses for a potentially transformative future, as socio-spatial interactions within the PCN remain constrained within the larger discursive *frame* (Stibbe 2015) of the City in Nature discourse. In Chapter 4: Ambivalent Discourses, Yock, Hannah and I identify the pitfalls of this ecotopic discourse of Singapore, such as its propensity for treating nature as a resource to proliferate from, and using the naturalization of spaces as a shallow solution to this, the root problem of developmental logic. Even in my analysis of *beneficial* discourses of the PCN, I will briefly touch on the *destructive* discourses perpetuated by this space in

Figure 5.1 Map of the Eastern Corridor section of the PCN. This is an example of the extensive connectivity of the network - such maps can be found at different sections along the various park connectors. For an overview of the network across the country, see the country-wide map of Singapore in Chapter 1. (Photo credit: Kristina.)

Figure 5.2 Photograph of Pasir Ris Park Connector in the Eastern Corridor of the PCN. (Photo credit: Kristina.)

the section *Network*. However, I address this by arguing for the PCN as a space that holds the potential of a future that has yet to be defined. Stibbe writes, 'Evaluations are stories in people's minds about whether an area of life is good or bad' (Stibbe 2015, 84). Through this evaluation according to the book's ecosophy,

I make a case for the present-day and future potentials of the PCN to reframe our relationship with the ecosystems of Singapore to be one that is based on non-hierarchical interactions that are defiant of and challenge the violent logics and ideologies of human domination over nature.

As with every space, plural discourses exist and are in relationship with each other. Especially since the PCN is still in development and to be finished in 2030, I turn to the PCN as 'text' to hone in on its potential to bring *beneficial* discourses into *salience* in Singapore. I frame my analysis of these discourses with the following questions: What discourses do the PCN facilitate, outside of top-down ascribed narratives? What communities and human and non-human energies surround the PCN? How can stories of *resistance* (Stibbe 2015) in the face of dominant *destructive* discourses arise from the PCN? The answers to these questions have important implications for how urban residents of Singapore navigate state-sanctioned land-use planning and our agency in facilitating more small-scale shifts in our relationship to Singapore's ecologies of alive places; towards one that 'value[s] and celebrate[s] the lives and wellbeing of all species, promote[s] human wellbeing, call[s] for reduction of consumption, and promote[s] redistribution of resources' (Stibbe 2015, 16).

Park: Forging Beneficial Relationships to Space, Time and Nature

In the island-wide map of Singapore that shows the PCN, it depicts the network as a web of park connectors that run through the entire country. These routes run through parks, alongside roads and along Singapore's '*longkangs*', or the re-naturalized canals that make up the drainage infrastructure of the city (NParks 2024). The connectors extend through a myriad of different environments, from 'predominantly natural – such as woodlands, unspoiled coastal shores and wetlands – to predominantly urban – such as city centers and industrial estates' (Briffett et al. 2004, 62). The PCN's web-like visual structure is an image of a viable alternative transport system to navigate the city-state, especially a way to travel without resorting to combustion engines that emit greenhouse gases and away from the danger, inconvenience and pollution of the main roads. It is also a way to travel that curates meeting spaces between humans and the non-human residents of Singapore on terms that are relatively neutral and even, thus lessening the omnipresent human domination of hyper-managed Singapore. While there is demarcation between the 'park connector' space and the 'urban sidewalk' space, in practice, residents of Singapore use them in interconnected ways – walking or cycling routes between places can switch from sidewalk to park connector, and back again. As such, park connectors in Singapore are used in everyday journeys and transits through the city.

Interestingly, this posits the PCN as a liminal space of transit – both as a way to get from one place to another, as an in-between experience of using the PCN as travel or recreation, and as a space that makes *salient* the more-than-human ecologies of alive places of Singapore. On top of its original purpose to be a space for exercise and relaxation, the PCN serves as a backdrop to people's daily living – in their commute across the city. Through this, Singapore is *reframed* (Stibbe 2015) from Housing Development Board (HDB) neighbourhoods dependent upon rail and road, to an interconnected route of non-vehicular mobility that invites Singaporeans out of their flats and cars and out into the more-than-human neighbourhoods of Singapore proper. The structure of its paved trails intertwining with the rest of the city allows for an alternate way of movement from place to place and supports our ecosophy's call for *Community*.

From this structural coding of the PCN emerges a discourse that has the potential to *resist* (Stibbe 2015) dominant discourses surrounding transport and movement through the city. A dominant narrative in Singapore about transport is one that prioritizes speed and efficiency. This is evident through (1) private transport channels, where multiple policy and infrastructural interventions work in tandem to reduce the number of cars on the roads and thus the amount of traffic; (2) public transport systems, such as the Mass Rapid Transit system and the bus system, which provide a robust, well-oiled machine that minimizes commute time as much as possible; (3) the pressures of vehicle ownership as a status of upper class incomes and having realized a key dream of petroculture: private car ownership. This discourse reflects a larger worldview and ideology of prioritizing productivity, which are key values underpinning hyper-capitalistic Singaporean society that prioritizes maximum capital accumulation in its national identity-formation and development story (see the introduction and Chapter 1 for more on this story, especially on growthism) (Freund 2010). David Harvey's concept of 'space-time compression' relates this shrinkage of spatiotemporal relations to past and present colonial logics of increased accumulation of labour and resources, as well (see introduction). These are the same ideologies that have resulted in extractivism (Chagnon et al. 2022) or the exploitation of resources (of natural resources; human labour; ecosystem health; time; and non-humans) for further economic gain and development. Hence, this dominant narrative around movement in Singapore is intrinsically linked to *destructive* discourses, the very ideologies our ecosophy hopes to transform.

Contrary to Singapore's prioritization of efficiency and speed, the PCN structurally *promotes* a discourse (and actual haptic and physical practice) of slowed transit. The park connector paths are spacious enough and smooth enough to allow for multiple activity-use – for cyclists, joggers, walkers, skaters and people in wheelchairs. The flora and vegetation lining the PCN paths also provide paths for fauna to transit through – both domesticated animals and wildlife. For example, many local forest and urban bird species flit through the trees and dragonflies, reptiles and otters can be found travelling along park connectors that run past *longkangs* too (NParks 2024). This space for movement through the city where diverse actors move alongside each other through non-motor means has *beneficial* implications on discourses around movement and time. Here, I reference literature surrounding velomobility, which offers possibilities of re-thinking mobility as convivial and social practice – one that is accessible, inclusive and encourages environmental connection (Horton 2006). As non-motor transit is slowed, it allows people to be more grounded in their experience of the environment and ecosystem around time. Elongating spatiotemporal processes allows for more space for authentic interactions with the ecosystem to occur. For example, it is easier to notice the people you pass by, a bird perched on a branch, or the rich and diverse flora breathing all around you. In his study of speed, Taylor (2014, 345) writes:

> Waiting bestows the gift of time, which is nothing less than life itself. When life slows down, it becomes possible to reflect thoughtfully on what usually rushes by too fast for us to notice… The faster we go the more we forget, and the more we forget, the less we know who we are or where we are going.

People who traverse the PCN are provided with an opportunity to experience an alternative relationship to space, time and nature – they are being re-peopled, re-humanized and re-natured with the slowed, more visible presence of the human and non-human in daily commute. The discourse promoted here is one that centres an embodied relationship with place and its ecosystem inhabitants. Every actor, including non-human ones, that partakes in this alternate relationship thus becomes more embedded in the socioecological fabric of the ecosystem within the PCN (Cox 2022). Through the network, transit opens up a new realm of play with space and community – both human and non-human! The *beneficial* discourse arising here can be summarized as such:

> People begin to redefine their relationship to space, time and nature within the city through interactions within the PCN. The 'green corridors' invoke a *re-minding*[3] of our embeddedness in the larger ecosystem – a discourse that is

> often erased by the isolation and alienation we feel in capitalist logics of space and time. We begin to feel a part of the *Community* as a whole, instead of alienated from other human and non-human actors in the ecosystem.

There are also lessons and implications here for the mindsets that govern our navigation of a future in the face of the climate crisis. As we work together across disciplines and sectors to create resilient responses to the ecological crisis, the urgency that the climate crisis demands often prompts a need for speed (Horton 2006). This discourse within the PCN also prompts us to view solutions borne from this with a *Critical* lens – often, the work of cultivating meaningful ecological identities and radically transforming the *stories-we-live-by* requires long-term attention and patience, work that spans generations and heals the past, present and future. This is not to say we must continue sitting in apathy and inactivity in the face of the climate crisis – but rather, we should reflect on which socio-economic infrastructures actually promote *beneficial* discourses of being in relationship to the ecosystem, instead of repeating the same logics that led us to this situation in the first place.

Connector: Interconnectivity within the System, Embedding the Self in the Whole

Beyond the types of movement that occur within the PCN, another physical quality of the network is its connecting linear pathways. From a structural perspective, the PCN embodies connecting nodes across the city. This interconnectivity is beneficial in multiple ways. From a biological perspective, the linear, interconnected habitat design of the PCN and the speeds and types of human movement it facilitates have been shown to be beneficial for attracting migratory and resident bird species, providing a viable habitat for bird residence (Briffett et al. 2004). When built with the conservation of habitats surrounding the network pathways in mind, a study showed that this contributed to the conservation of bird species within the area (Sodhi et al. 1999). From a structural perspective, the interconnectivity of the park connectors within the PCN builds up a *salience pattern* of the image of connecting nodes across the city-state – instead of viewing different spaces in silos. This promotes a message of linkage across spaces in Singapore and the dependence of these different natural spaces on each other.

Furthermore, along the PCN, there are signboards that can be found at various junctures on the paths that aim to signpost the presence of diverse actors within the PCN's ecosystem, and that signal the need for consideration towards other users of the space. For example, there are signs that remind people not to litter to avoid pollution and to not harm flora, fauna and other people. There are also signs for cyclists to slow down at certain junctures of the PCN that might have more heavy foot traffic, or steeper sloping paths. These signs are *salience patterns* (Stibbe 2015) that recur throughout the PCN and bring the more-than-human-world further into the minds of people who experience that PCN. This is a reminder of the PCN's interconnectivity across diverse actors – bringing to *salience* an awareness of embeddedness of the self within a larger network and ecosystem.

This interconnectivity – both across habitats in Singapore and across the human and non-human – promotes an awareness of the interdependency that can exist within an ecosystem. Stibbe writes that such *salience patterns* counter 'the abstraction and erasure of the more-than-human world in so many texts that we come across in everyday life' (Stibbe 2015, 177). By repeatedly seeing these signs, actors navigating the PCN network are constantly reminded that their actions do not occur in silos, rather that they affect other aspects of the ecosystem. Instead of positioning nature as an untouchable, pristine resource that should be conserved through separation from the human (see Chapter 3 on dualisms), nature is now a part of our lives in meaningful ways.

This can be corroborated with other studies of urban residents' biophilic encounters within the PCN. Briffet et al.'s (2004) study assessed Singapore residents' relationship to nature through their experience of the park connectors and green corridors of Singapore. Most respondents reported a positive association with nature such as an enjoyment expressed towards being in contact with flora and fauna, especially due to their everyday proximity to nature through the PCN. The participants of the survey also connected personal well-being to proximity to nature through the PCN. More face-to-face, embodied interactions with the natural world, especially in urban areas, can lead to more ecocentric perspectives and support for actions that move society towards more just and sustainable futures (Taylor et al. 2019). Such data offers support towards our ecosophy and why the PCN is a *beneficial* text for Singapore: it helps promote the ideologies of *Community* and also *Care*.

In fact, in recent years, this ecosocial ecosystem has also translated to digital platforms. There exists a Facebook forum called 'SG PCN Cyclist'[4] created in 2012, which is a platform for cyclists to 'share your favourite PCN paths and photos in this group'. Cyclists who travel through the PCN share their encounters with others in the ecosystem (both human and non-human) and swap stories about the PCN. Some examples of biophilic and community-focused statements from such cyclists include:

- 'Started our morning embracing mother nature and cherishing the friends in our riding journey. Thank you all that have joined and I'm honoured to be part of the team. Hope to see you guys again on our next ride'. Clement See, June 5 2023
- 'Parks and nature are places of respite. Let's be a part of this move to fight against climate change'. Nur Hasanah Osman, April 8 2023
- 'Extreme weather. Accelerated climate change by human activities is real. We are bearing the brunt of what the previous generation gave us. Let's not do the same to our next. We, the citizens, can do what we can on our end. Hopefully the relevant authorities and big corporations can take their role seriously too. Lead the way, we'll follow. We can't be hippies these days anyway. #justride' Zhu Bingcheng, January 2 2021
- 'A last minute riding session until wee hrs in Rail corridor Friday night. Enjoying the total darkness and silence of the nature. Though many times nearly blinded by the bright white flash lights from oncoming cyclists. finally did a pit stop to refuel at Rail Mall Springleaf Prata. I want to thank the 2 tables of guys (whom I guess also cycling along Rail Corridor) who so nice n quickly offered their tables to us when he saw I was so kan cheong (translation: urgent) looking for seats. Whoever you are and If you chance upon this post, Thank you!! Though I might have forgotten to thank you guys yesterday – at The Rail Corridor. There many nice and friendly cyclists around and with good company make the whole cycling experience such a joy'. Maydianne Yip, October 23 2021.

Already, from this brief look into urban residents' responses to the PCN, there are shifts in urban residents' relationships to the ecosystem within the PCN to be one of *Care*. As explained in Chapter 2, *Care* is to recognize that 'to be one is to become with many'. A *beneficial* discourse arises and can be summarized as:

> I am a part of a larger *Community* – the ecosystem as a whole! My actions have an impact on others and their actions impact me as well. Hence, I must manage these relationships within the *Community* with *Care* for the other as I would extend to myself, such that everyone in the ecosystem can live lives of wellbeing and dignity.

The above discourse has radical implications for futures of sustainability in Singapore. As mentioned in the introduction and Chapter 1, a key *destructive* discourse in Singapore is alienation: of human from non-human; of self from ecosystem; of the country from regional relationships surrounding climate vulnerability and destruction. If more in Singapore were to understand the importance of extending *Care* to others in our shared, global ecosystem, this could transform our relationships not just with nature, but with others in the global network of vulnerability to the polycrisis. This discourse within the PCN promotes a community-centric lens to Singaporean worldviews, which could facilitate the *salience* of well-being for all to be prioritized in wider urban and social imaginaries.

Network: Building Together, Challenging Hierarchy within the Ecosystem

From the ecolinguistics analysis above, I have examined how the PCN is a space coded for the potentiality of sublime experiences with the ecosystem through the cadence of its socio-spatial interactions, and the *salience* of interconnectivity as a recurring theme throughout the space.

However, it must be acknowledged that the PCN maintains a discursive positionality and complicity within the larger frame of Singapore's ecomanagerialism. Here, I reference Chapter 4's critique of the treatment of (1) the country's ecotopic narrative as an *erasure* of underlying power imbalance and dualism between human and nature (see the City in Nature case study analysis) and (2) the use of nature to reify and uphold the dominant structure of subjugation by providing a balm to our lived realities under capitalism, numbing us from its effects and distracting us from challenging these dynamics (see also the reference to Zizek in Chapter 4's Botanic Gardens case study). This is a cause for concern, especially considering the upscaling of the PCN across Singapore such that all households are slated to be within a ten-minute walk from a park by 2030, and the hopes for its implementation in other geographical contexts (Wang et al. 2019). The PCN is complicit in concealing some socioecological harms, such as: the destruction of primary forest in the name of development that has occurred throughout Singapore's history; the continued harms implicated in conservation strategies as colonial construct;[5] the PCN's complicity in furthering the acceptability to justify ecosystem damage by replacing it with managed vegetation in urban green spaces; and the employment of foreign labour as park cleaners that are responsible for the upkeep and maintenance of these green spaces, who often have to work late hours within a larger national labour system where there remains a lack of appropriate labour policy for these workers in precarious positions.[6] These continue to uphold the unequal power dynamic between the human and non-human that is a foundational ideology of Singapore.

How can we treat the PCN space with a *Critical* lens, honing in on the root of our damaging relationship to the ecosystem – the hierarchy that exists in unjust relationships with those on the margins of the ecosystem (human and non-human)? The *Critical* aspect of this book's ecosophy, which calls for 'a constant and critical re-assessment of our positionality and the invisible exclusions and hierarchies that might be embedded in our imagining of a better world', prompts us to further question the negotiations of power within the PCN's textual discourses, and whether they are truly *beneficial* and strive for justice across the ecosystem.

In light of this, I argue that while the PCN can produce *destructive* discourses such as those outlined above, these *destructive* discourses largely stem from top-down imposed narratives – the *beneficial* discourses that can emerge occur through unmanaged, anarchistic interactions that occur within ground-up relational dynamics among actors in the system are the ones championed in this chapter's interpretation of the 'text' of the PCN. These subaltern interactions have more potential for transformative resistance to the hierarchies of imbalanced power that reinforce the structures of subjugation that govern urban residents' relationship to the ecosystems of Singapore. Here, I reference Llewyellyn et al. (2022), who portray ecosystem emergence as a bottom-up, collective discovery and process of negotiation to tease out the *beneficial* discourses that can arise from the PCN.

A potential *beneficial* discourse that can emerge in this way is inclusion and equity, as a way of eroding the human–nature binary. Another *salience pattern* built up in the PCN is its recurring presence and ubiquity throughout the city-state, and its quality of openness – as a free-of-charge, un-gated space, it promotes access to different groups of people – regardless of age, class, gender and race – and it promotes both access to and interaction with the non-human in unscripted and unmanaged ways. With such access, actors within the ecosystem not only share the PCN space but also experience *Care* from and within the PCN space in various ways.

First, the trees and shrubbery provide access to respite from urban heat for climate-vulnerable communities in Singapore: people who exercise can exercise in shaded areas, reducing risk of heat stroke; shade for delivery riders on bicycles, who often use various pathways within the PCN to get to their delivery locations; elderly, children and differently abled people also similarly benefit from the cooling effect of the natural features of the PCN. As a country one-and-a-half degrees north of the equator, one of the most pertinent climate effects we are vulnerable to is rising temperatures. Hence, it is important to have infrastructures within urban infrastructure that account for heat vulnerability and that also contribute to reducing the urban heat island effect (Zaki et al. 2020). The PCN is also a space that provides mobility and access: the routes are wheelchair-friendly through gently sloping bridges and smooth roads in park connectors to access nature spaces, and non-human others of Singapore also use the space to traverse the island at their own pace and forms of mobility.

Second, the PCN's transitory nature as a passage route between places has also made it a space that is often used as a third space, or space for interactions outside productivity or capital exchange to occur; in short, the PCN can function as a form of peripheral commons, both for the human and non-human alike (Soja n.d.). This is also due to the absence of overt enclosure of the space – unlike some other natural spaces in Singapore, the PCN is open at all times of the day with no lifeguards or security guards to enclose the space. On the weekends, people hold picnics, sing karaoke, schedule parties and gatherings and exercise in these pockets of open space along the PCN that become sites of interaction and relationship-building. Especially for queered populations (Oswin 2014) such as migrant construction workers and foreign domestic workers in Singapore, these spaces can be a respite from work and can provide opportunities for congregation with friends in Singapore. Here, I reference Oswin's (2014) extension of the term 'queer' to include the figure of the foreign worker in Singapore, a community that is subject to precarity and marginality in Singapore due to their transient presence and inability to fit into the heteronormative logics of the 'responsibly reproductive subject' (Oswin 2014, 423) that govern spatial power within the city-state. Furthermore, spaces of inclusion and rest in Singapore are lacking for such historically marginalized communities (Yeoh and Lam 2022). To see the PCN open up possibilities of redefining how space is used and who has access to these spaces stimulates hopeful visions for a future Singapore that prioritizes the well-being of all who reside in the city-state and resists exploitative relationships towards communities on the margins.

Third, relationship-building in non-hierarchical ways can occur through the messy, unplanned and unmanaged interactions that occur daily in the PCN. Interactions with nature have become 'transgressions' – this is evidenced by frequent news articles in Singapore's mainstream news outlets reporting on sightings of wildlife, positioning these instances as peculiar and notable. The PCN normalizes these interactions as wildlife traverse through the space as well – when an otter family runs across a bike path, or a monitor lizard creeps up the side of the *longkang*. A sort of anarchist calisthenics (Scott 2012) is exercised here, in defying the socialized 'laws' of separation between human and nature in Singapore. Through these frequent exposures, residents become more and more accustomed to being a part of nature and integrated in a shared ecosystem with the other. Thus, this *beneficial* discourse arises:

> *Community* should be expanded to include *all* in the ecosystem, even those historically alienated through dualisms – the non-human and those on the margins. These relationships should be defined through dynamics of *Care*, instead of hierarchy.

This discourse challenges the binaries and hierarchies that enforce the current world order of polycrisis, including in Singapore. As we continue to live lives of embeddedness in our interactions with the ecosystem within the PCN, we can further hone the skill of questioning *Critical*-ly and contemplating power dynamics. The *Critical* approach of the book's ecosophy helps us to identify harmful dynamics within socioecological relationships more readily, pivoting us towards just futures.

Majulah…?: Moving Onward in the PCN

With the articulation of the *beneficial* discourses that are enmeshed in physical spaces in Singapore, such as the PCN, we can understand how they conversely construct our own interactions with nature. In fact, depending on which discourses we pay attention to, they play an important role in constructing the futures of our relationships with nature.

In analysing the PCN, I call to attention the complexity of its discursive quality as a space. While it has potential to be analysed in a *beneficial* way such as I have done in this chapter, I am hesitant to end this case study overly-abstracting the possibilities the PCN offers. While these *beneficial* discourses can exist, they remain on the margins of our hegemonic relationship to the ecosystem in Singapore. The reality is that despite the presence of *beneficial*, non-hierarchical and regenerative relationships to nature and to each other, the ecosystem remains abstracted in most Singapore residents' minds and concealed in our everyday lived realities.

The purpose of this writing, tied to the larger aim of the book and the power I and the co-authors find in our ecolinguistic analysis of Singapore, is to highlight the potential for these marginal discourses to provide a path of *resistance* against the dominant *stories-we-live-by* – specifically those that reinforce an exploitative and extractive relationship to the ecosystem. Through amplifying discourses of *resistance* and our agency in uplifting *beneficial* discourses that can already exist in our city, I hope that stories like those within the PCN can empower more of us in Singapore to take ownership of rewriting our Climate Generation's story and future.

Bukit Brown: Blending Nature, Culture and Heritage

Beverley Choo Jia Ying

Singapore's natural landscape has changed drastically over the years. Documenting this change in his chapter in *Nature Contained: Environmental Histories of Singapore*, Tony O'Dempsey (2014, 45), writes: 'During the early years [of independence], very little consideration was given to retention of any remaining natural habitat, no matter if it was secondary forest or regenerating mangrove habitat. Industrial and residential development was considered a "do or die" activity'. This phenomenon is exemplified well by various spaces, including Pulau Bukom which was discussed in Chapter 3 where the government displaced local communities and completely deforested the island to transform it into a massive fossil fuels and petrochemicals complex. As time went on though, the government's approach transformed and it began to prioritize some level of nature conservation. This is reflected in the state's aim to 'transform itself into a City in Nature by 2030' (Ministry of Sustainability and Environment 2024). While no longer *destructive*, this discourse as analysed by our ecosophy in Chapter 4 is still *ambivalent* at best[7] as the government continues to pursue industrial and infrastructural development at the cost of Singapore's few remaining natural spaces. Since 1819, Singapore has lost almost forty per cent of its biodiversity, and just one measly per cent of our original primary forests remain today (Esterman 2023).

One of the more recently contested spaces in Singapore at the time of writing has been Bukit Brown Cemetery. Originally home to an estimated 100,000 graves belonging to Singapore's early Chinese immigrant population (Ng and Lim 2024), Bukit Brown Cemetery (colloquially referred to as just 'Bukit Brown') was opened on 1 January 1922. It was Singapore's first Chinese municipal cemetery, where Chinese folks of various clans and dialect groups were, for the first time, buried in the same cemetery ('Bukit Brown Cemetery' n.d.). The last burial took place on 30 November 1972, and when the cemetery closed burials at the start of

1973, it became the largest cemetery in Singapore for the war dead in situ (Leow and Lim 2016, 6–12). Since its establishment, the vegetation in Bukit Brown Cemetery has grown and flourished to form a lush secondary rainforest. Today, it is home to not just the dead, but a myriad of species of flora, fungi and fauna, including two families of long-tailed macaques.

The Bukit Brown area has been earmarked for residential development since the Urban Redevelopment Authority's (URA) Concept Plan 1991, a long-term strategic land-use and transportation plan (Urban Redevelopment Authority 1991). But the threats to its existence have only really materialized in the early 2010s. By analysing the space, the various discourses surrounding it, as well as the many efforts to conserve it using *salience* and *re-minding* techniques, we find that the cemetery is a site of immense potential for Singaporeans to connect with each other and with the more-than-human.

Lornie Highway and Developing Bukit Brown

For decades, little public attention was paid to the cemetery, except for the occasional visitors and families paying respects to their ancestors. This is not to say that the cemetery was abandoned – during Qing Ming,[8] the cemetery would become so filled with people that traffic advisories have been published to advise motorists to avoid the main road that leads to the cemetery, Lornie Road (*Today* 2010). Bukit Brown was only shot into the forefront of Singaporean's imaginations in May 2011, when the government announced that they decided to start 'developing' the Bukit Brown Cemetery area, beginning with the construction of a highway that would affect a portion of the cemetery space. The cemetery was to be bifurcated, and thousands of graves were to be exhumed for the construction of the eight-lane Lornie Highway.

In response to a letter from a member of the public questioning the need to destroy Bukit Brown Cemetery given the availability and greater suitability of other plots of land in Singapore for housing development (Chan 2011), the URA's Director for Physical Planning at the time, Mr Tan See Nin, wrote, 'The URA has been consciously conserving both built and natural heritage in our planning for Singapore. For instance, just across the road from Bukit Brown is our Central Catchment, a large protected tropical nature reserve which has a special place in the heart of many nature lovers'. Here, Tan reveals an *ideology*[9] that alleges that: (1) natural spaces are only important to a small and niche group of people (i.e. 'nature lovers') and (2) natural spaces possess only instrumental and not inherent or intrinsic value. On the latter point, Tan suggests in his response that it is acceptable to lose the natural space at Bukit Brown since Singapore already has another natural space, the Central Catchment, which should be sufficient to meet 'nature lovers' desire for green spaces. Underlying this *ideology* is the failure to understand the *unique* value of each natural space, its distinct ecosystems and distinct individuals of various species that call it home. This *ideology* enables the government to frame their decisions to destroy natural spaces for the purposes of 'development' as utilitarian and generative, when the preservation of these spaces could be just as, if not more, utilitarian and generative (think reducing the urban heat island effect [Shishegar 2014], improving air quality [Jaung et al. 2020] and providing space for recreation for the many that use the cemetery's paths as their running and dog-walking routes).

Indeed, this story told by Tan and the government in general has, unfortunately, become one of Singaporeans' *stories-we-live-by* and generated an *erasure pattern*[10] in local discourse about conserving natural spaces. Opinion pieces written by members of the public about the government's announcement to build Lornie Highway in the Singaporean newspaper *The Straits Times* included quotes like, 'The Bukit Brown habitat can easily be recreated' (Wee 2015) and 'We cannot afford to be too sentimental over preserving an old cemetery like Bukit Brown, which is today nothing more than a forlorn place with broken tombstones and overgrown *lalang*'[11] (Heng 2015). Yet, a quick trip to the cemetery will show that such nonchalance and flippancy are truly unwarranted (Figure 5.3).

I first visited Bukit Brown Cemetery in 2020 and immediately fell in love with the space. I was enamoured by the sheer biodiversity of plants I could see, the artistry of the tombs and the stories that the space held. While many sections of the cemetery are paved in gravel, reaching certain other tombs requires muddying your shoes and manoeuvering through dense foliage. Some graves have been overtaken by various species of moss, ferns and creepers, and others overcome by the natural process of soil creep have become slanted and no longer stand upright. Conversely, some of the tombs are very well-maintained, surrounded by uniformly cut grass and decorated with fresh flowers and fresh paint over the tombstone engravings. Still, that doesn't stop the macaques from using them as stepping stools and lunchtime spots! As articulated in Chapter 2, under the *Community* pillar of this book's ecosophy, we highlight how 'we want to identify spaces that promote the integration and co-existence of diverse actors and plural ways of being, and enable the right of all to thrive and to live lives of well-being and dignity'. Bukit Brown epitomizes such a space, bringing together not just humans and non-humans, but non-living humans and non-humans too, as descendants pay respects to their ancestors to the backdrop of birdsong and mulch under their feet.

Figure 5.3 Graves surrounded by lush greenery at Bukit Brown Cemetery. (Photo credit: Beverley.)

Unlike Singapore's perfectly manicured and government-managed parks and gardens,[12] the flora at Bukit Brown can be unruly and wild. Even where it is managed, it is often conducted in a bottom-up manner, rather than a top-down one. While the National Environment Agency's horticultural team maintains the trees along the edges of the cemetery that border roads, the vegetation inside the cemetery is maintained by three tomb keepers and the descendants of those buried there. Descendants are able to use their own tools or engage contractors to trim and cut the vegetation surrounding their ancestor's graves as they see fit. In this manner, Bukit Brown resonates with our ecosophy's *Critical* pillar, as it challenges the *de facto* relationship that Singaporeans have with greenery – instead of being a passive observer, we are empowered to become active participants who engage directly with vegetation. Whether it is in the form of cutting down a tree that stands in the way of your ancestor's grave or simply having to remove the fallen leaves stuck to the bottom of your shoes at the end of your visit, the cemetery invites more intimate ways of engaging with the non-human.

Ultimately though, Lornie Highway was constructed and opened on 19 April 2019 (Land Transport Authority 2019) at the cost of vast deforestation and the exhumation of over 4,000 graves (Ng 2024). Justifying the decision, a spokesperson of the URA told the press that, '[P]lanning for the long term in land-scarce Singapore does require us to make difficult trade-off decisions' (Han 2013). However, it seems that this decision cannot possibly be *that* difficult, since the government almost always chooses infrastructural development over the preservation of natural and heritage sites. This is not a new phenomenon either. A newspaper article published in 1971 about the government's takeover of cemetery land made an acute observation: 'Our dead are sleeping directly on the path of progress and development' (Tan 1971).

In hypermodern Singapore, this 'path of progress and development' is unrelenting. Yet, Bukit Brown stands as a direct challenge to the growthism and productivism that dominate Singaporean society.[13] The space invites us to slow down and redirect our attention away from the hustle and bustle of life in the most fatigued country in the world (Chua 2024) – it calls us to remember our ancestors, engage with cultural and religious rituals, learn about and reflect on our history, and truly, take a second to stop and smell the flowers (including those of the seventy-six nationally threatened plant species that can be found at the cemetery (Jong 2022)). Although the government decided to move forward with the partial destruction of the cemetery and still plans to eliminate the rest of it in due time, many civil society groups in Singapore also see the value of Bukit Brown and the *beneficial* discourse it represents.

Saving Bukit Brown

When the government announced plans to build the highway in 2011, several grassroots efforts that engaged in what ecolinguists have termed *re-minding*[14] and *building salience*[15] emerged. Some of these efforts included a petition initiated by SOS Bukit Brown, all things Bukit Brown [sic], which began giving free guided tours of the cemetery, and Nature Society Singapore and Singapore Heritage Society, which both published position papers that emphasized the unique value of the cemetery and engaged the press to spread the word. For the purposes of this book, we'll dive deeper into atBB's efforts since their *re-minding* and *salience-building* efforts are substantial, ongoing, and fit into our book's larger call for strong sustainability and degrowth in urban cities given climate trajectories.

Shortly after the government's announcement was made, brothers Raymond and Charles Goh organized various tours of the cemetery for interested members of the public. From there, some attendees became more curious and interested in the space – so much so that they attended multiple tours and eventually became volunteer guides themselves. These passionate tour attendees-turned-guides later became known as 'the brownies', who organize under an informal initiative called 'all things Bukit Brown' (atBB). The brownies didn't just organize tours, though. In an attempt to conserve the cemetery and call attention to its cultural significance, the brownies submitted an application to the World Monuments Watch to have Bukit Brown Cemetery listed as a Watch Site. Their application was accepted, and in 2014, Bukit Brown became Singapore's first WMF Watch Site. In the same year, the brownies were recognized as the 'Advocacy Organisation of the Year' in the inaugural Singapore Advocacy Awards; 'Arguably, the brownies' efforts have succeeded in cultivating something more precious than the cemetery itself – a growing community of Singaporeans who treasure their common heritage and are determined to keep it alive' (Leow and Lim 2016, 11–12).

Indeed, what the Singapore Advocacy Awards were recognizing was the brownies' efforts in *re-minding* and building *salience* around Bukit Brown Cemetery, and its ecological, cultural and heritage value. They did this in multiple ways. The first and most well-known of such efforts is their tours of the cemetery. The brownies have been conducting guided walking tours of Bukit Brown Cemetery for over a decade now, guiding over 27,000 people to the site at time of writing.[16] During these tours, the guides share stories of the people buried there, the stories of the graves themselves and more.

The brownie who guided a tour I attended in 2020, for example, showcased tombs of a number of prominent and wealthy individuals that are beautifully decorated with intricate carvings, imported hand-painted tiles and accompanied by sculptures and statues. In particular, he pointed out the grave belonging to Cheang Hong Lim, a philanthropist and Chinese community leader whose name continues to live on through the well-known Hong Lim Park – the only venue where public protests can legally take place in Singapore. He also highlighted the grave of Oh Sian Guan, a founding director of Singapore's oldest bank, the Oversea-Chinese Banking Corporation (OCBC Bank), and great-grandfather of Kevin Kwan, the author of the book *Crazy Rich Asians*, which was later adapted as a widely popular film distributed by Warner Bros. Pictures. By calling to mind such familiar names, the brownies build on existing *salient* ideas from our daily lives, from pop culture to living advocacy spaces, and incorporate the cemetery into it (Figure 5.4).

Besides tours, the brownies also organize community *re-minding* events. For example, on 18 August 2024, the brownies organized the Bukit Brown Spirit Film Festival. The four-hour event comprised screenings of four short documentaries,

Figure 5.4 Sculptures and artwork at Bukit Brown Cemetery. [Left] One of two Sikh guard statues at the tomb of Chinese physician and former Qing official Chew Geok Leong (d. 1932). Sikh men who took up employment as security guards for private employers became so well-trusted and iconic that they were enshrined in stone, replacing statues of Chinese warrior gods and guardian animals who traditionally protected the dead from harm. There are approximately thirty such statues at Bukit Brown Cemetery (Aggarwal 2017). [Right] Sculpted into the stone walls of this tomb are pictorial depictions of Chinese fables. Being so close to the ground, these miniature single-frame stories are at the perfect height for children to see and thus learn from when their families come to visit and pay their respects to the deceased. (Photos credit: Beverley.)

a panel discussion with the filmmakers, and a debut presentation and tasting of 'Blend B', a Chinese tea curated for Bukit Brown by Pek Sin Choon, a homegrown Nanyang tea merchant established in 1925. The documentaries screened explored spirituality and the afterlife – Chinese burial and exhumation rituals in particular – in a rapidly modernizing, land-scarce Singapore. Through these films, the brownies call attention to the transformation and loss of traditional rituals, and with it, our connection to our ancestors and kin. That is, they are *re-minding* us of the importance of kinship and connection through shared tradition and history, as facilitated by spaces like Bukit Brown Cemetery. By screening so many films produced by demographically diverse filmmakers, atBB also points to the *salience* around the cemetery that has been building slowly over the years since the government's decision to build Lornie Highway. This is not a forgotten space, nor is it one to be forgotten.

Ultimately, through guided walks and events like the film festival, the brownies call our attention to the imminent *erasure* of Bukit Brown Cemetery and its role in preserving history, maintaining kinship ties through rituals and more. Through this *re-minding*, they not only advocate for the conservation of the cemetery itself but also bring *salience* to the ways in which we connect with each other – as well as with non-humans and the non-living – and how these methods are changing as our landscape changes. The brownies effectively highlight how Bukit Brown embodies this book's ecosophy that advocates for a Singapore that replaces the constant pursuit of growth with the pursuit of happiness and flourishing for all in our community (comprising humans and non-humans alike).

The Future of Bukit Brown

> We have always believed Bukit Brown as a landscape of memory markers that can lead you to places in the past you have never been, and into a future you can now begin to reimagine.
>
> - Claire and Catherine, co-founders of all things Bukit Brown
> (Leow and Lim 2016, 17)

In August 2024, atBB and Singapore Heritage Society opened the memorial called 'Sounds Of The Earth', a collaborative outdoor installation designed by artist Chen Sai Hua Kuan using unclaimed artefacts recovered from the graves exhumed to make way for Lornie Highway. The memorial sits between the highway and the cemetery's graves and vegetation that were spared… for now. Much like the cemetery itself, the lifespan of the Sounds Of The Earth memorial is limited. At the time of writing, it exists on a three-year temporary occupation license, with the possibility of no more than two three-year extensions thereafter. Space next to the cemetery has already been used to build the foundations of the Bukit Brown train station as part of Singapore's wider Mass Rapid Transit (MRT) system – likely in preparation for the eventual transformation (destruction and reconstruction) of the area to become a residential zone. To have such a one-of-a-kind space that could become a cornerstone of Singapore's heritage be placed in such a position of precarity is a real shame.

A popular idea in the Environmental Humanities is the act and phenomenon of 'becoming-with'. Attributed to the influential scholar Dr. Donna Haraway, the notion of becoming-with challenges anthropocentrism and emphasizes how humans and non-humans are entangled and interdependent (Haraway 2010, 53–5). Bukit Brown Cemetery is a place where every visitor becomes – with the ancestors to whom they pay their respects, the birds whose songs fill the air, the vines and mud that cling to their shoes and the fellow visitors with whom they share a smile as they cross paths.

How Bukit Brown currently functions as a 'text', then, is undeniably as a form of *beneficial discourse* according to our ecosophy. The space represents the real possibility of cohabitation and interconnectedness between humans and other (non-)living beings around us. By going beyond merely considering relationships between living beings, the cemetery and its advocates expand our ecosophy and urge us to consider our relationships between the living and non-living too. It calls attention to the role that history, culture and heritage play in how we understand *Community*, enact *Care* and practice healthy *Criticism* in the context of a country that seems to be constantly in pursuit of growth and development in spite of the climate crisis and the sixth mass extinction.

Chek Jawa: Singapore's Pocket of Coastal Treasures

Hannah Ng Shueh Yi

An Island (not so) Far Away: Pulau Ubin

The journey to Chek Jawa Wetlands is long – as with most other journeys to marine gems in Singapore. The ferry alone is a fifteen minute ride, dependent on weather and the boat uncle,[17] not yet considering the hours it could take to travel to the ferry point, located at one of the northeasternmost tips of Singapore's mainland. Upon disembarkation at the only public jetty[18] on Pulau Ubin, Chek Jawa is either a 3-kilometre hike, bicycle ride, or a short van ride further. In addition to the journey, Chek Jawa's full features are only visible at tides below 0.5 metres in height, which, despite the diurnal tides in Singapore, is surprisingly difficult to catch during daylight hours. Even then, the tidal visitation window is usually less than two hours and at that only approximately five times a month. Visitation, therefore, requires a dedicated trip – chance encounters are difficult if not impossible for most mainlanders. Like many port-turned-megacities, Singapore's coastlines have been primarily converted and dominated by human use into recreational, industrial, or residential spaces. Singapore's coastal spaces, which have retained natural ecological features, are rare, and Pulau Ubin contains most of them.

Pulau Ubin, one of Singapore's northeastern offshore islands, is a known rustic getaway for nature lovers and peace-seekers, maintaining an unparalleled cultural and ecological space in ultra-modernized Singapore. Ubin has retained some *kampongs* and their inhabitants, deserted quarries, hiking trails and unrivaled biodiversity. Its landscape includes four quarries, hills and lower-lying coastal areas, including Chek Jawa Wetlands, extensive mangroves and coastal forests. Scattered through the landscape are public and private campsites, trails and cemeteries. Several born and bred islanders continue to reside on Ubin permanently, though many islanders were relocated in anticipation of redevelopment in the 1970s–2000s (Ang 2014). NParks estimates that Pulau Ubin is also home to over 786 native plant species and over 500 animal species, including birds, butterflies, mammals, reptiles and amphibians, many of which are very rare, or not found on mainland Singapore at all (National Parks Board 2021). Apart from rare flora and fauna, Ubin is also home to the Chek Jawa Wetlands whose conservation-based discourse is the topic of this *beneficial* ecolinguistical case study.

Tanjung Chek Jawa, Chek Jawa Wetlands, or simply Chek Jawa,[19] contains the meeting point between six types of coastal and marine habitats: a seagrass lagoon, a rocky shore, a sandy beach, coral rubble, coastal and mangrove forests. In 2001, when conservationists first 'discovered' it, the unique intersections between these coastal habitats were likened to finding 'a family heirloom' (Lim 2001).

Saving Chek Jawa, 2001: Mobilizing Collective Action[20]

In 2001, conservationists from Singapore's mainland 'discovered' Chek Jawa's ecological richness and catapulted it into the spotlight. *Remember Chek Jawa* (2007), a documentary by Eric Lin, featured Joseph Lai recounting his astonishment at Chek Jawa's existence and his accidental discovery of it:

> In January 2001, I volunteered [as a nature] guide ... [We went] to see the forest in Chek Jawa, [as] I am a botanist by training... Because it was low tide then, by chance, we found the sandflat. Instead of going to the forest, we went [out on the intertidal] to see. And we [saw] so many wonderful things. That was the first time we saw such a wonderful array of animals there.[21] (Lin 2007)

His joy was short-lived, for he soon learned that Chek Jawa and the rest of Ubin were slated for reclamation and redevelopment in early 2002 – villagers were already being resettled in anticipation of the reclamation. As explored in Chapter 1, land reclamation is an important feature of Singapore's developmental narrative, informing many uniquely Singaporean discourses of pragmatism. The

Urban Redevelopment Agency (URA) and the HDB, statutory boards under the Ministry of National Development (MND), were the primary agencies involved in the impending redevelopment of Pulau Ubin.[22] Thoroughly convinced Singapore could not afford to lose Chek Jawa, Joseph Lai went to a public consultation session with the URA to appeal against the planned reclamation, and he began an effort to bring public and state attention to the incredible biodiversity Chek Jawa contained.

This kickstarted Chek Jawa's spotlight in the media, with newspaper articles in Mandarin and then English-based press beginning in July 2001. Meanwhile, the Nature Society of Singapore (NSS), Raffles Museum of Biodiversity Research (RMBR – now known as the Lee Kong Chian Natural History Museum), and other volunteers began an urgent effort to visit, catalogue and guide the public to visit Chek Jawa before permanently losing the site to reclamation. This consisted of intertidal biological and geomorphological surveys – requiring a suitably low tide date and a small army of volunteers; as well as online outreach and logistical support to bring members of the public to visit. The scientific effort consisted of a muddy group of around forty volunteers using guides to identify species, recording their presence with photographs and collecting voucher specimens for the museum. The geomorphological study consisted of another group of volunteers led by coastal geomorphologist Teh Tiong Sa. Many volunteers were barely prepared for the mud, sand and stinging organisms in the intertidal zone – some ended up barefoot with their shoes so full of mud they had fallen off.[23] Ecological data they collected would later be submitted by experts to the URA as proof of ecological value in consideration of their appeal against reclamation in Chek Jawa. Another group, including now local veteran conservationist N. Sivasothi (also known as Otterman) (N. Sivasothi, n.d.) started blogs and self-organized nature walks for the public to visit and experience Chek Jawa for what they expected would be the last time in October of 2001. Volunteers created makeshift stations on Chek Jawa to guide[24] the public to appreciate the organisms and landscapes on which they found themselves. On a single weekend in October 2001, over 1,000 visitors made the journey, young and old, to say hello and goodbye (Teh and Raju 2010). All these efforts inspired filmmaker Eric Lin to follow the efforts and make the documentary titled *Remember Chek Jawa* (2007) where he was intrigued by the public's 'efforts to participate in a tiring, muddy and apparently incomprehensible biodiversity survey, at a time when all hope had virtually been lost'(Teh and Raju 2010, 10).

Throughout, the URA maintained that 'Chek Jawa['s] reclamation [was] decided after careful study', as the study found it unsuitable for coral reefs or maintaining resident dugong[25] populations and therefore would not have a significant impact (*The Straits Times* 2001). The significance of Chek Jawa was the point of contention between nature groups, and an extension of the public, and the developing agencies. Checklists of species were important ecological evidence countering HDB's study and calling attention to biodiversity outside charismatic species and habitats (dugongs and coral reefs). The conservationists involved documented and accounted for Chek Jawa's intertidal and coastal habitats and their biodiversity, habitats that HDB's study had not considered. Singapore's policy and decision-making rely heavily on scientific evaluation and pragmatism to inform land-use decisions, using studies pre-construction to ensure impacts to natural environments are mitigated – thus, quantitative scientific evidence was a valuable tool for Chek Jawa's discourse.

The construction of Chek Jawa's *beneficial* discourse relied not only on scientific means to promote Chek Jawa. It also included a select group of conservationists coming together and using social means to promote Chek Jawa, aiming to share its natural beauty with the wider public. These groups called for volunteers, who came from all walks of life – 'these were people with a big heart' (interview with Sivasothi, *The Straits Times* 2001). They came together to preserve and commemorate pieces of Chek Jawa, sharing knowledge of it freely through guides, aiding in transporting people to Chek Jawa and creating routes to bring members of the public through the intertidal while minimizing their impact on the non-human life in Chek Jawa. The cause of Chek Jawa created stronger conservationist *identities* in the community of volunteers and visitors involved in these initial stages. Unlike broader *identities* where others ascribe features of *identities* (Stibbe 2015), many people involved in Chek Jawa's conservation took on a *self-identity* as protectors, researchers, collectors and preservers, as seen for example, in these anecdotes from volunteers:

- Now I know, more than before that in Singapore, a country which many people say is famed for apathetic people, but if you have a project that you believe is worth doing, and if you put out a call, I believe you will find people who will respond to it. (Volunteer; Eric Lin You Wei 2007)
- Smelly, muddy job! But someone has to do it. (Unidentified volunteer; Eric Lin You Wei 2007).

Linguistically, these excerpts reflect people's individual and collective responses, or an answering – Chek Jawa provided an opportunity for volunteers to come together, experience nature in a novel way and connect that effort with an overarching conservation discourse. The online and in-print media presence of Chek Jawa further reinforced these *identities* by creating access to the space by uploading photographs of landscapes and their contained biodiversity, thus making the place more *salient* for all Singaporeans, thus challenging the anthropocentrism and growthism of governmental ideology.

These identities required a level of *Critical* analysis (in the re-imagination of Singapore they wanted), challenging the broader, dominant narrative of 'development' (as per Neo 2007) through scientific and social *framings*. Importantly, they found legal and socially acceptable means to be *Critical*; using participatory modes of advocacy based on science and *Community*. However, advocates still operated under competing discourses, as their identities were still rooted in the dominant Singapore story as a core *story-they-live-by*, as they acknowledged the difficulty of balancing developmental needs specific to land-scarce Singapore and the preservation of space (Interview with Joseph Lai, Eric Lin You Wei 2007).

Chek Jawa Online and In-Print

A key text in Chek Jawa's discourse is Chek Jawa's blog (chekjawa.nus.edu.sg, maintained by Sivasothi). This blog was instrumental in documenting ecological findings and improving accessibility to Chek Jawa in publishing public guided walk dates and other news related to and concerning Chek Jawa (interview with Sivasothi 2024). The website continues to be an important identification and archival resource for aspiring ecologists and nature enthusiasts on Chek Jawa's floral and faunal presence (Figure 5.5).

Other than Sivasothi, nature groups contributed to the Chek Jawa discourse through articles, advocacy and research. For example, NSS's Nature Watch called Chek Jawa 'A Treasure to Behold' with 'sponges galore'; Chek Jawa was hailed as 'the Model Teacher, a natural outdoor classroom' (S. C. Chua and Lum 2001, Title; Lum 2001, Title). Shawn Lum, then President of Nature Society Singapore wrote, 'I can vouch for the fact that students who visit the world of Chek Jawa do not walk away without being touched by the sublime beauty of the place, the magnificence of the flora and fauna there, and a sense of amazement that such places exist in Singapore' (Lum 2001).

Figure 5.5 Chek Jawa online. (Photo credit: Hannah.)
Screenshot of the virtual tour with landscape photos of different parts of Chek Jawa. Virtual tours catered to visitors who could/did not make the journey to Ubin. Photos taken from the archives of chekjawa.nus.edu.sg. (Screenshot credit: Hannah.)

In *Chek Jawa: Discovering Singapore's Biodiversity*, Chua Ee Kiam called Chek Jawa 'a showcase for the miracle of life and life is a gift from Heaven' (Chua 2002, Abstract). The framing of Chek Jawa as 'heavenly' or 'Treasure', and the linguistically positively coded words describing the human–nature relationship with it ('amazement', 'magnificence') all served to create the legacy of Chek Jawa's preservation story, helping to make the ecosystem *salient* and *re-minding* Singaporeans of Chek Jawa's value. As a *multimodal discourse*, the chekjawa.nus site also contributed to shaping the scientific, visual and intertextuality (between print media, online media and reality) that is now coherent with Chek Jawa's conservation discourse. These *appraisal patterns* reinforced positive emotional associations between Singaporeans (broadly) and Chek Jawa, cyclically pervading the conscious and subconscious worldviews and ideologies of Singaporeans surrounding Chek Jawa then and now. The continual *re-minding* of Chek Jawa's beauty and biological rarity, especially of its more-than-human lives as important and present in seemingly human-dominated Singapore, used *salience patterns* to entrench the worth of Chek Jawa and other comparably ecologically rich areas in the minds of Singaporeans. The 2007 documentary further distinguished Chek Jawa by resisting homogenizing ecological spaces for its audiences (Lin 2007). By entangling the unique visual images of the organisms and landscapes of Chek Jawa with the volunteers' voices, videos and narratives, Chek Jawa's discourse of conservation, *Care* and *Community* was emphasized. This documentary *foregrounds* Chek Jawa's specific and unique identity – which continues to be watched, shared and made *salient* in the minds of Singaporeans today.

The *Community* that rallied around Chek Jawa transformed it from a space of *erasure* to a space of *salience*. The dominant discourse of Chek Jawa was created by the actions and inclusion of over 1,000 members of the public, nature lovers, filmmakers, scientists and others who dedicated time and effort to protecting, remembering and loving this space. Despite their diverse backgrounds and demographics, their collective narrative coalesced, creating and expanding the *Community* of Chek Jawa both in the inclusion of different social circles and in the introduction and inclusion of non-human communities to more people. While retaining their individual agency and identity, they were also able to represent themselves on many occasions as a collective – 'over a thousand people', 'nature groups', 'volunteers' and 'people'. This is reflected in both the eventual outcomes and linguistic patterns of narratives. Volunteers and visitors were brought closer to Chek Jawa's non-human nature through the language of science, ecology and 'can-do' attitudes.

An important caveat is that the assessment of nature through inherently abstracted methods of quantification continues to perpetuate a relationship of quantitative basis instead of focusing on the inherent and intrinsic value of a space/an ecosystem. Such quantification distances humans from seeing and experiencing Chek Jawa as an ecology of an alive place, full of non-human others and the unfolding of life on its own terms, instead focusing on quantifying, rarifying and up or downscaling (Isti'anah, Suhandarno and Winarti 2024). This is an important area of improvement, as post-conservation Chek Jawa discourse should move beyond scientific or quantitative validation and now consider as valid other forms of qualitative discourses that foreground and make *salient* such non-human life. Rather, Chek Jawa's *salience* can now be *framed* and understood through the space's inherent value as home to non-human nature in the *framing* of ecological richness, and equally, its history of traditional ecological knowledge, home to Indigenous practice (explored below), its aesthetic and therapeutic value, its cultural heritage values and its ability to provide climate resilience to Singapore. Chek Jawa can therefore provide a microcosm 'snapshot' of how Singapore can reframe the *stories-it-lives-by*, leading to the type of *beneficial* discourses and sustainable practices it will need to navigate the twenty-first century.

Despite these areas of discursive improvement, Chek Jawa's discourse left a twenty-three-year legacy that continues to shape Chek Jawa as a positive, *beneficial* discourse that is testament to the dedication of the conservationists of 2001. The connection and relationship built between the public, certain community stakeholders and Chek Jawa created such an uproar (a *disruptive* discourse!) that in January 2002, the Ministry of National Development announced the deferment of the reclamation of Chek Jawa for ten years. In this stage of independent Singapore's history, appeals for conservation had rarely succeeded – the only prior success had been Sungei Buloh Wetland Nature Reserve in 1993 (and around the same time, the gazetting of Labrador Nature Reserve in 2001). Chek Jawa and the larger Ubin's conservation was a win, even if only for a decade.

Chek Jawa Reclamation Deferred: 2002 – Onwards

Chek Jawa continues to be a beacon of Singaporean biodiversity, with the interface of habitats creating the perfect conditions for abundant life forms within the small space as it contains a plethora of marine organisms and species. The socioecological conditions of Chek Jawa in 2001 also continue to be a dominant story of conservation in Singapore – other than the physical visitors in 2001, the 2007 documentary and subsequent news features, books and research that have been completed to date continue to cultivate the identity of Chek Jawa beyond its ecological beauty. Chek Jawa's story became an important stepping stone in the current environmental advocacy scene Singaporeans continue to enjoy today – many nature lovers in Singapore continue to reference the websites started and popularized because of Chek Jawa (Sivasothi 2001–2007).

However, since 2001, Chek Jawa has experienced some serious changes due to society, state and climate. In 2007–2008, teams monitoring Chek Jawa noticed massive die-offs; large carpet anemones, seagrass and crabs suddenly disappeared after a heavy period of rainfall with freshwater flux changing the ecological composition of Chek Jawa (Loh, Sivasothi and Todd 2009).[26] After the mass die-offs, the construction of two boardwalks, the viewing tower, the viewing jetty and the visitors centre were undertaken in a 6.2 million dollar project (Precise Development Pte Ltd. 2023) with a goal to minimize human impact while allowing high visitorship.

As a discourse, however, Chek Jawa's narrative of successful advocacy via scientific methods and public engagement continued to inform other conservation successes in Singapore. In 2008, local marine academics published the Singapore Blue Plan based on their expert knowledge. This document details recommended steps for protecting, preserving and balancing Singapore's marine spaces and organisms with Singapore's relentless pace of development (Hamid et al. 2009). It included the gazetting of a small marine park around Sister's Islands – which today is Singapore's only protected marine space. This effort was renewed by another generation of marine academics in 2018, with further support and inclusion from more stakeholders, and submitted for government consideration.[27] In both marine and non-marine spaces, nature groups recognize the power of civil environmental society, and many[28] continue to work with the state system of public and private consultations to raise issues regarding development, conservation and hopes for better managing Singapore's remaining wild spaces.

In identifying *beneficial* discourses for this book, Chek Jawa stood out to me because of its important history of inspiring social and scientific advocacy for protecting parts of the last remaining coastal habitats of Singapore. This aligns with our ecosophy as Chek Jawa embodies an expansion of *Care* and *Community*, as well as being *Critical* by advocating for non-human nature against development – its protection and public accessibility fulfil all three pillars of our ecosophy in various ways. Its effects on subsequent conservation efforts in Singapore, the continued maintenance of the space as a lasting beacon of conservation, and the care and community it has cultivated in Singaporeans make Chek Jawa a significant space for all of Singapore.

Barriers and Displacement: Chek Jawa's Ambivalent Discourse

As the space has evolved over the subsequent two decades of deferment, it should be no surprise that the Chek Jawa of today contains many differences. Though the main framing of Chek Jawa is *beneficial*, it is important to acknowledge the *destructive* or *ambivalent* aspects of Chek Jawa and how these offer caution to what is overall a relatively *beneficial* discourse of Singapore.

For one, the boardwalks keep visitors up and away from the actual intertidal zone. The only full visitation experience is mediated by guided tours, either permitted by or run by NParks. These restrictions were put into place after Chek Jawa was 'loved to death' during the high visitorship after the events of 2001, a protective measure for non-human life who are easily stomped on by humans. However, the physical distance and mediation of the space by the boardwalk create a barrier to each visitor's relationship with the space. Like in many other parks in Singapore, the boardwalk was the compromise between organisms and habitat versus allowing visitors an up-close and personal experience, aiming to increase accessibility while minimizing the direct impacts of individuals on Chek Jawa's delicate life forms. While the ecology is better preserved, it is more difficult for people to *Care* for spaces they cannot fully encounter, and therefore, whose interrelation is lower in quality (Adams 2007). The ability to feel, touch and connect with Chek Jawa and its varied habitats is reduced, and the opportunity for interrelationship (Buber, Kaufmann and Buber 1996) is reduced. Not only is the individual visitor's ability to interrelate reduced, but the story of nature as *abstract-ecological-distanced* is enacted in the minds of Singaporeans. By creating a distance between us and Chek Jawa, we have inexplicably distanced ourselves, thus creating relationships that are anthropogenically mediated, shielded and restricted by the structure of the boardwalk. How can anyone fully experience Chek Jawa Wetlands without getting their feet wet? Considering this, Chek Jawa shifts from *beneficial* to *ambivalent* discourse, highlighting the difficulty of creating truly *beneficial* discourses, let alone maintaining them (although from the point of view of the mudflats and its organisms, the boardwalk is a *beneficial* text as it protects their well-being).

Another *ambivalent* factor of Chek Jawa is seen in how the government of Singapore has treated the original inhabitants of the island. For example, Nor

Syazwan bin Abdul Majid, known for his blog *Wan's Ubin Journal*, documents and advocates for the *Orang Pulau* community, the indigenous islanders of Pulau Ubin and the other islands of Singapore. His social media accounts and blog are a rich repository of island-descendant perspectives documenting historical islander life from his mother's islander family and community while offering reflections of a deep personal solastalgia. For his mother, Chek Jawa was a foraging, '*berkarang*', landscape for intertidal creatures. Wan shares how her feet got wet and sandy on Chek Jawa's shores while harvesting carpet anemones, which were widely used in curry, *sambal* and *rendang* – staple hearty communal foods indigenous to the island's culinary dishes of the 1900s.[29]

Wan's writings and online content sharing[30] are full of longing. The poems he writes about his ancestry and the island are filled with 'Oh' and 'Can you'; discourse markers coupled with a lyrical questioning structured into an ask for forgiveness from Chek Jawa personified as Wan's distant grandparent (see Figure 5.6). His yearning to reconnect with the spaces his mother knew and his grief for the loss of place-based culture and understanding that Singaporeans of my generation suffer are emotions not unfamiliar to me – my grandmother's Singapore is vastly different from mine. His writings emphasize that Chek Jawa, in particular, has always been special, holding different socioecological meanings, purposes and relationships to different people through time.[31] *Wan's Ubin Journal*, though at times seemingly nostalgic, provides another avenue of critical analysis of Chek Jawa – for him and his community Chek Jawa has become a space of dispossession from cultural practice and indigenous dwelling. This is the biggest *destructive* aspect of Chek Jawa's history and space and is a common discourse across most of Singapore's offshore islands and its coastal areas.[32] Many of the Ubin Orang Pulau community living next to Chek Jawa were instructed to re-settle in 2001 for the upcoming 2002 reclamation – their homes were promptly demolished except for House No. 1, a British bungalow. After the Save Chek Jawa movement in 2001, Ubin Orang Pulau watched bittersweetly as their home, though saved from reclamation, was instead fenced up and gazetted so no further fishing or foraging could take place. This particular narrative needs little analysis. The *erasure* of culture and practice is negative in any context, especially ones of dispossession for development. Wan and other islander descendants continue to experience a solastalgia that is all too common within Singapore's rush towards ecomodernity. It is also a reality often *erased* from the Singapore Story and from Singapore's reputation as a 'green' city of the future.

Chek Jawa Today: Peeking through the Boardwalk

Today, Chek Jawa remains one of the more 'wild' places in Singapore where nature is *framed* as central, and human-made urban ecologies are largely *erased*. Chek Jawa's aesthetic, scientific and holistically health-inducing space appeals to many urban dwellers of Singapore. This appreciation and care for Chek Jawa is apparent even during restricted high tide views. The expanse of sea and sky is a stark contrast to the high-traffic, high-rise-dominated Singapore mainland – a respite from the man-made streets and unending buildings in the city's labyrinth. However, as of 2024, Chek Jawa is mostly accessible only by boardwalk with signage instructing visitors to refrain from venturing off track, with railings and suspension at levels that deter one from venturing onto the majority of Chek Jawa. Instead, the accessible parts of Chek Jawa include a restored Tudor-style house containing explanations and accounts of the history of Chek Jawa; a viewing jetty; two boardwalks going through the six major habitats; and a viewing tower allowing visitors a peek into what the wetlands offer.

The two boardwalks, Coastal and Mangrove, respectively, take visitors through the six habitats. The coral rubble and seagrass lagoon are completely submerged at high tide, and the seagrass lagoon shows only as shadowy shapes underwater. Scattered along the sandy beach are washed-up seagrass mounds that fill the air

Figure 5.6 *Wan's Ubin Journal* instagram posts and poem about Chek Jawa. (Screenshot credit: Hannah.)

with pungent saltiness. The Coastal boardwalk leads into the Mangrove boardwalk, where an extensive mangrove forest clicks and sputters nearby. At the junction between the start of the mangroves and the end of the sandy beach, hundreds of fiddler crabs wave their claws at me. The mangrove forest grows along a different stretch of coast, preferring muddy areas which are regularly submerged. Propagation efforts have left structures with intermediate mangroves growing out of them. Deeper in the mangrove forest, I am surrounded by mud mounds, the homes of mud lobsters that bury deep and create mounds of up to three metres high by the base of mangrove trees. Fresh mud dribbles out of certain mounds, traces of the elusive creature. Through and around the mounds are small estuarine streams, home to insects and other smaller organisms, fish and the occasional mud crabs.

A giant mudskipper flops right by the boardwalk beneath my feet. We stare at each other for a second, I am in awe to see it. It seems unbothered by my presence, and I sit in an encounter.

Chek Jawa of the Future

It is important to acknowledge that in Singapore, as is the case in the USA, Canada, Australia and other forcefully acquired territories, we have Indigenous populations whose losses and sacrifices have provided the land, ecology, beauty, space and time upon which settlers built modern nations and societies. For ***Majulah...?*** to be fulfilled, we must embrace the multiplicity of traditional practices that communities like Ubin Orang Pulau keep alive. *Beneficial* discourses are the discourses we want to continue to grow, which can lay the foundations of the future. Chek Jawa's discourse is a *story-we-live-by* that testifies to Singapore's community and their dedication to ***Majulah...?***, serving as a beacon of hope and hopes for the future. Its foundations have taken root in the past two decades, and continuing to strive to improve, grow and learn from it can hopefully continue to entrench this *beneficial* discourse about conservation and diverse ways to value Singapore's coastal areas in the minds of Singaporeans.

To continue to fully realize Singapore's socioecology and for strong sustainability, Singapore needs to prioritize not just green spaces but also the brown and blue ecologies of the sea, the mud and everything in between. At the time of publication, Chek Jawa and Pulau Ubin are still considered 'Reserve Land', deferred for reclamation and redevelopment until otherwise decided. The precarity of this place, and of the indigenous islanders who once called it home, are sadly still present even in the most *beneficial* discourses and places of Singapore. Rather than a perfect and essentialized *beneficial* discourse, the mobilization to save Chek Jawa suggests that Stibbe may be correct when he theorizes that, 'The aim of the form of ecolinguistics described in this book [and the one we are using] is to encourage people to be *more* critical and reject or accept discourses based not just on personal or social considerations but ecological ones too' (Stibbe 2015, 16, italics in original).

To be truly *beneficial* and to fully align with our ecosophy, we invite Singaporeans to be even more critical of how their society treats the indigenous peoples of its islands. We encourage Singaporeans to work with the government and other stakeholders to permanently protect Chek Jawa while allowing for its indigenous peoples to return to steward the island in strongly sustainable ways, thus adding to Singapore's vibrancy as a society while increasing climate and ecological resilience. Though practices may have faded with time, there can be space for renewal of tradition and expansion of *Care* and *Community* in the *Onward* march of Singapore (Figure 5.7).

Conclusion

Hannah Ng Shueh Yi

> The aim is to contribute to efforts to raise awareness of – and resist – ecologically destructive stories, and promote uses of language which tell different stories that encourage people to protect the systems that life depends on (Stibbe 2015, 16).

Through the explorations of the three case studies here, each representing a different aspect of Singapore from daily life on the Park Connector Network, to sites of ecological and cultural significance like Bukit Brown and Chek Jawa, we have highlighted the aspects of Singapore that we most want to see in our *Onward* future. From the interconnectedness between self, space and nature, to the historical collective power that shaped and continues to shape Chek Jawa, we have highlighted specifically and uniquely Singaporean spaces and 'texts' that embody the being of both one and many (Haraway 2015) – of humans and non-humans. The Park Connector Network embodies an exemplar of a *vocabulary* of interconnectivity drawn between human and non-human, and a burgeoning *beneficial* relationship to time, space and nature. Bukit Brown offers an example of how direct engagement of culture, heritage and beyond-humans (both non-living humans and non-humans) can inspire *Community* and deepen eco-centrism. Chek Jawa is one of many conservation wins in Singapore – many

Figure 5.7 Photos from Chek Jawa. (Photo credit: Hannah.)

other cases (small pockets of treasures) are scattered across the island and whose discourses may be no less *beneficial* than Chek Jawa's. *Promoting* these stories is key to creating the Singapore we hope to see, a resilient, inclusive, caring and critical Singapore of the future. These three case studies underline the strengths of ground-up and highly localized formation of *Community*; their strengths come from creating specific contexts of *Care*, which come from the unique attributes of each 'text' and the way *beneficial* discourses have *Critically* been shaped around them.

Equally important is that we also highlighted some areas of improvement, for these discourses are not perfect and have areas that must be *resisted*. As Stibbe elabourates, the final stage of a *beneficial* discourse is *promotion* – not of the material, text, or space itself, but of the '*way of writing or speaking that tells a useful story*… what is promoted is a specific cluster of linguistic features (pronoun use, grammatical structure, presuppositions, positioning or participants, etc.) which tells a particular story' (2015, 33, italics added by us). The focus is thus on the values being *promoted* in each case study – of interwovenness, appreciation and advocacy. As we conclude this book, we encourage readers to reflect on the ways in which these discourses may reflect in their own lives and how to *promote* or *resist* their life-giving or life-taking ideologies.

Key Questions

1. What is the dominant *vocabulary* of our daily lives? What are the spaces which espouse these values, and how can we shape these spaces to be more *beneficial*?

2. Given Singapore's reliance on science and evidence-based policy decisions, how can future conservation efforts better incorporate the inherent value of ecologies and honour Indigenous peoples and practices? Is there a better balance between these considerations and those of land scarcity, population growth and climate change?
3. How can each and every person espouse more *beneficial* discourses in our daily lives? Which of the case studies most resonates with our lived realities?

Notes

1 See https://normandychairforpeace.org/lines-of-research/good-stories/.

2 Here, I am referencing the collection of personal essays in *Making Kin: Ecofeminist Essays from Singapore*, which writes to the concept of 'kinship' as experienced by women in Singapore who navigate 'overlapping discourses of family, home, ecology and nation' (Vincent and Poon 2021, 18). This book was pivotal in shaping an ecofeminist lens to the concept of *Care* and *Community* in our ecosophy (see Chapter 2), specifically to the geopolitical context of Singapore.

3 Stibbe coins the term *re-minding* to describe this discursive phenomenon, where signs such as the above explicitly call attention to the embeddedness of all actors within the PCN as part of an interconnected and interrelated *Community*.

4 The Facebook forum can be found as a public community group here: https://www.facebook.com/groups/SGPCN/.

5 See Chapter 4: Ambivalent Discourses, specifically the Botanic Gardens case study on Singapore's current green landscape and its roots in colonial history.

6 This is further expanded on in 'Coming Clean: A Study on the Well-being of Bangladeshi Conservancy Workers in Singapore', a report written by Singapore-based non-governmental organization called Humanitarian Organization for Migration Economics (HOME).

7 See Chapter 4 for an in-depth analysis of the 'city in nature' discourse.

8 Qing Ming is a festival observed mainly by those of Chinese descent for the remembrance of one's ancestors.

9 Defined by Stibbe (2015, 23) as 'belief systems about how the world was, is, will be or should be which are shared by members of particular groups in society'.

10 Defined by Stibbe (2015, 146) as 'a linguistic representation of an area of life as irrelevant, marginal or unimportant through its systematic absence, backgrounding or distortion in texts'.

11 *Lalang* is the regional name for *Imperata cylindrica*, a tall grass native to Southeast Asia.

12 See Chapter 4 for an in-depth case study on the Singapore Botanic Gardens.

13 See this book's introduction and the introduction to this chapter for more in-depth discussion about these topics.

14 Defined by Stibbe (2015, 162) as 'explicitly calling attention to the erasure of an important area of life in a particular text or discourse and demanding that it be brought back into consideration'.

15 Defined by Stibbe (2015, 162) as 'a story in people's minds that an area of life is important or worthy of attention'.

16 This is anecdotal and comes from an interview with Catherine Lim, co-founder of all things Bukit Brown.

17 Boat uncle's refer to the captains of the boats.

18 There are actually four jetties on Pulau Ubin, but the others are private-access or abandoned.

19 Chek Jawa is thought to refer to 'Javanese Men's Point', as many Javanese men went to Pulau Ubin to harvest granite. The rocks of Ubin share the same geological history as a third of the Singapore mainland – termed Bukit Timah Granite (1994). These same rocks form the rocky shores of Chek Jawa and are the rocks for which Ubin was named (Majid 2023; Ang 2014).

20 Recounted while referencing interview with N. Sivasothi, August n.d.; Navjot Sodhi Lecture 2024; Chek Jawa websites chekjawa.nus; Eric Lin You Wei 2007.

21 I would love to catalog and describe in detail the many organisms that Joseph and the other nature-lovers found. However, for brevity, this lies outside the scope of this case study.

22 Though we wished to further explore land reclamation case studies, it unfortunately lay outside the scope of this book.

23 Any experienced guide of the intertidal knows that many stinging, sharp, and otherwise uncomfortable sensations can await you from the various marine organisms living in the intertidal. Twenty three years prior, perhaps this was less well-understood by those from the mainland.

24 Less experienced volunteers were given a blurb that they learned on the go to share about the area they were stationed at.

25 Dugongs are a native large marine mammal, locally considered critically endangered.

26 The subsequent life of Chek Jawa is omitted for brevity. Knowledge of local ecosystems is primarily accessed by academia or via wildlife blogs.

27 See Jafar et al. (2018)

28 At the point of writing, at least nine well-known active nature groups are known to attend consultations on a semi-regular and relevance basis (personal experience, 2023–2024).

29 Modifications of similar dishes continue to be staples in many Malay and Peranakan communities in Singapore and in Straits Chinese communities.

30 I have been lucky to have heard him speak, and many involved with this book project have received a heritage walk from Wan and Todd included such a walk in subsequent offerings of his Foundations of Environmental Humanities course at YNC.

31 Orang Laut and Orang Pulau groups (which directly translate to Island People or Sea People) are accepted to be the indigenous populations from pre-colonial times. Their descendants who continue to advocate for preservation of their cultural heritage include Wan and oranglaut.sg (descendents of the islanders of Pulau Semakau). Their important stories unfortunately lie outside the scope of this book but can be found on their respective blogs, social media and other articles.

32 Other known groups include Orang Seletar and Orang Biduanda Kallang of the Seletar River and Kallang River, respectively.

References

Adams, Will. 2007. 'The Primacy of Interrelating: Practicing Ecological Psychology with Buber, Levinas, and Merleau-Ponty'. *Journal of Phenomenological Psychology* 38 (1): 24–61. https://doi.org/10.1163/156916207X190238.

Aggarwal, Vandana. 2017. 'The Sikh Guards of Bukit Brown'. *The New Paper*. https://tnp.straitstimes.com/news/singapore/sikh-guards-bukit-brown.

Ang, Seow Leng. 2014. 'Standing Firm: Stories of Ubin'. *Biblioasia*. National Library, Singapore. https://biblioasia.nlb.gov.sg/vol-10/issue-2/jul-sep-2014/stories-of-ubin/.

Briffett, Clive, Navjot Sodhi, Belinda Yuen, and Lily Kong. 2004. 'Green Corridors and the Quality of Urban Life in Singapore'. In *Proceedings 4th International Urban Wildlife Symposium*, 56–63.

Buber, Martin, Walter Arnold Kaufmann, and Martin Buber. 1996. *I and Thou*. New York: Simon and Schuster.

n.d. 'Bukit Brown Cemetery'. *Roots.Gov.Sg*. Accessed 7 September 2024. https://www.roots.gov.sg/resources-landing/online-exhibitions/bukit-brown-cemetery.

Chagnon, Christopher W., et al. 2022. 'From Extractivism to Global Extractivism: The Evolution of an Organizing Concept'. *The Journal of Peasant Studies* 49 (4): 760–92. https://doi.org/10.1080/03066150.2022.2069015.

Chan, Ronald. 2011. 'Why Destroy Natural Habitat When Other Land Available?' *Today*, 15 June 2011. https://eresources.nlb.gov.sg/newspapers/digitised/article/today20110615-2.2.26.3.

Chua, Ee Kiam. 2002. *Chek Jawa: Discovering Singapore's Biodiversity*. Singapore: Simply Green.

Chua, Ilyda. 2024. 'S'pore Is the Most Fatigued Country in the World. We Ask S'poreans from 3 Generations What They Do to Stay Alive'. *Mothership.sg*. https://mothership.sg/2024/01/fatigued-spore-world-alive/.

Chua, Sek Chuan, and Lum, Shawn. 2001. 'Pulau Ubin's Tanjung Chek Jawa a Treasure to Behold'. *Nature Watch*. https://habitatnews.nus.edu.sg/pub/naturewatch/text/a093a.htm.

Cox, Peter. 2022. 'Vélomobility Is to Degrowth as Automobility Is to Growth: Prefigurative Cycling Imaginaries'. *Applied Mobilities* 8 (3): 265–85. https://doi.org/10.1080/23800127.2022.2087134.

Esterman, Isabel. 2023. 'Study: Singapore Biodiversity Loss Is Bad – But Not as Bad as Previous Estimate'. *Mongabay Environmental News*. https://news.mongabay.com/2023/12/study-singapore-biodiversity-loss-is-bad-but-not-as-bad-as-previous-estimate/.

Freund, Peter. 2010. 'Capitalism, Time-Space, Environment, and Human Well-Being: Envisioning Ecosocialist Temporality and Spatiality'. *Capitalism Nature Socialism* 21 (2): 112–21. https://doi.org/10.1080/10455752.2010.489684.

Hamid, Farid Abdul, et al. 2009. 'The Singapore Blue Plan 2009'. [insert publisher] http://nss.preview.orion5.iwi.com.sg/old/pdf/blueplan-final.pdf.

Han, Kirsten. 2013. 'Singapore: The Fight to Save Bukit Brown'. *The Diplomat*, 30 October 2013. http://thediplomat.com/2013/10/30/singapore-the-fight-to-save-buki t-brown/.

Haraway, Donna. 2010. 'When Species Meet: Staying with the Trouble'. *Environment and Planning D Society and Space* 28 (1): 53–5. https://doi.org/10.1068/d2706wsh.

Haraway, Donna. 2015. 'Anthropocene, Capitalocene, Plantationocene, Chthulucene: Making Kin'. *Environmental Humanities* 6 (1): 159–65.

Heng, Cho Choon. 2015. 'Bukit Brown not Worthy of World Heritage Status'. *The Straits Times*, 11 July 2015. https://www.straitstimes.com/forum/letters-in-print/bukit-brown-not-worthy-of-world-heritage-status.

Horton, Dave. 2006. 'Environmentalism and the Bicycle'. *Environmental Politics* 15 (1): 41–58. https://doi.org/10.1080/09644010500418712.

Isti'anah, Arina, Suhandano and Daru Winarti. 2024. 'Quantifying Nature in Tourism Discourse: A Corpus-Ecolinguistic Perspective'. *Jurnal Arbitrer* 11 (2): 172–85. https://doi.org/10.25077/ar.11.2.172-185.2024.

Jaafar, Zeehan, Danwei Huang, Jani Thuaibah Isa Tanzil, Yan Xiang Ow and Nicholas Yap, eds. 2018. *The Singapore Blue Plan 2018*. Singapore: Singapore Institute of Biology.

Jaung, Wanggi, et al. 2020. 'Temperature and Air Pollution Reductions by Urban Green Spaces are Highly Valued in a Tropical City-State'. *Urban Forestry and Urban Greening* 55: 126827.

Jong, Ying Wei. 2022. 'A Checklist of the Vascular Plant Flora of Bukit Brown and Lau Sua, and Suggestions for Improving Reporting Practices in Biodiversity

Impact Assessments'. *Nature in Singapore* 1: 172–89. https://doi.org/10.26107/NIS-2022-0119.

Leow, Claire, and Catherine Lim. 2016. *World War II @ Bukit Brown*. Singapore: Ethos Books.

Lim, Lydia. 2001. 'Pulling Together to Turn the Tide for Chek Jawa'. *The Straits Times*, 29 December 2001. https://chekjawa.nus.edu.sg/articles/defer/14.htm.

Lin, Eric You Wei, dir. 2007. *Remember Chek Jawa*. Documentary. https://youtu.be/EJI8Hr22-OE?si=jt6iVy9nFr7Vqep9.

Loh, K. S., N. Sivasothi and P. A. Todd. 2009. 'Mass Mortality and Recruitment of Intertidal Marine Invertebrates at Chek Jawa, Singapore'. In *Proceedings of the 14th NUROP Congress, National University of Singapore* 18: 4.

LTA. 2019. 'New Road Openings to Support Future Traffic Demand'. *Land Transport Authority*. Accessed 17 October 2024. https://www.lta.gov.sg/content/ltagov/en/newsroom/2019/3/2/new-road-openings-to-support-future-traffic-demand.html.

Lum, Shawn. 2001. 'Chek Jawa: Nature as Model Teacher'. *Nature Watch*. https://habitatnews.nus.edu.sg/pub/naturewatch/text/a093c.htm.

Majid, Nor Syazwan bin Abdul. 2023. 'Pulau Ubin'. *Wan's Ubin Journal*. 18 January 2023. https://www.wansubinjournal.com/pulauubin.

Ministry of Sustainability and Environment. 2024. 'City in Nature'. *Singapore Green Plan 2030*. 3 December 2024.

National Parks Board. 2021. 'Flora and Fauna of Ubin'. *National Parks Board: Pulau Ubin*. 10 February 2021. https://www.nparks.gov.sg/pulau-ubin/biodiversity/fauna-and-flora-of-ubin.

National Parks Board. n.d. 'NParks Park Connector Network'. Accessed 27 December 2024. https://pcn.nparks.gov.sg/.

Neo, Harvey. 2007. 'Challenging the Developmental State: Nature Conservation in Singapore'. *Asia Pacific Viewpoint* 48 (2): 186–99. https://doi.org/10.1111/j.1467-8373.2007.00340.x.

Ng, Keng Gene. 2024. 'New Outdoor Display at Bukit Brown Cemetery to Open by August'. *The Straits Times*, 8 July 2024. https://www.straitstimes.com/singapore/new-outdoor-display-at-bukit-brown-cemetery-to-open-by-august.

Ng, Keng Gene, and Kevin Lim. 2024. 'Markers of the Past: Unclaimed Tombstones at Bukit Brown Become Art'. *The Straits Times*, 28 August 2024. https://www.straitstimes.com/multimedia/graphics/2024/08/bukit-brown-cemetery-singapore/index.html?shell.

O'Dempsey, Tony. 2014. 'Singapore's Changing Landscape since c. 1800'. In *Nature Contained: Environmental Histories of Singapore*, 17–48. Singapore: NUS Press. https://doi.org/10.2307/j.ctv1nth2d.7.

Oswin, Natalie. 2014. 'Queer Time in Global City Singapore: Neoliberal Futures and the 'Freedom to Love." *Sexualities* 17 (4): 412–33. https://doi.org/10.1177/1363460714524765.

Oxford English Dictionary, s.v. 'connector (n.)', July 2023. https://doi.org/10.1093/OED/9170035308.

Oxford English Dictionary, s.v. 'park (n.)', December 2024. https://doi.org/10.1093/OED/2605619451.

Precise Development Pte Ltd. 2023. 'Chek Jawa'. Precise Development Pte Ltd. 2023. https://precise.sg/chek-jawa/.

Scott, James C. 2012. *Two Cheers for Anarchism: Six Easy Pieces on Autonomy, Dignity, and Meaningful Work and Play*. Princeton, NJ: Princeton University Press.

Shishegar, Nastaran. 2014. 'The Impacts of Green Areas on Mitigating Urban Heat Island Effect: A Review'. *The International Journal of Environmental Sustainability* 9 (1): 119.

Sivasothi, N. n.d. 'About'. *Otterman Speaks*. https://otterman.wordpress.com/about/.

Sodhi, Navjot S., Clive Briffett, Lily Kong, and Belinda Yuen. 1999. 'Bird Use of Linear Areas of a Tropical City: Implications for Park Connector Design and Management'. *Landscape and Urban Planning* 45 (2–3): 123–30. https://doi.org/10.1016/s0169-2046(99)00028-6.

Soja, Edward W. n.d. 'Thirdspace: Toward a New Consciousness of Space and Spatiality'. In *Communicating in the Third Space*. London: Routledge.

Stibbe, Arran. 2015. *Ecolinguistics: Language, Ecology and the Stories We Live By*. London: Routledge.

Tan, Abby. 1971. 'Tombstones Make Way for Big Progress'. *New Nation*, 3 June 1971. https://eresources.nlb.gov.sg/newspapers/digitised/article/newnation19710603-1.2.44.3.

Taylor, Bron, Jen Wright and Todd LeVasseur. 2019. 'Dark Green Humility: Religious, Psychological, and Affective Attributes of Proenvironmental Behaviors'. *Journal of Environmental Studies and Sciences* 10 (1): 41–56. https://doi.org/10.1007/s13412-019-00578-5.

Taylor, M. C. 2014. *Speed Limits: Where Time Went and Why We Have So Little Left*. New Haven: Yale University Press.

Teh, Tiong Sa, and K. Raju. 2010. 'Conserving Marine Biodiversity and the Role of Individuals: A Case Study of Chek Jawa, Pulau Ubin, Singapore'. Presented at the Growth, Equity and Environment: Missing Link between Academic Research and Policy Matters in Marine Park Areas, FEA, UM, Kuala Lumpur, April 15. http://www.globalislands.net/userfiles/Singapore_4.pdf.

The Straits Times. 2001. 'Chek Jawa Reclamation Decided after Careful Study'. July 27, 2001.

Today. 2010. 'Qing Ming Traffic Arrangements'. 19 March 2010. https://eresources.nlb.gov.sg/newspapers/digitised/article/today20100319-2.2.9.13.2.

Thomas, Llewellyn D.W., Erkko Autio, and David M. Gann. 2022. 'Processes of Ecosystem Emergence'. *Technovation* 115: 102441. https://doi.org/10.1016/j.technovation.2021.102441.

Urban Redevelopment Authority. 1991. *Living the Next Lap: Towards a Tropical City of Excellence*. Urban Redevelopment Authority. Call no.: RSING 307.36095957 LIV.

Vincent, Esther, and Angelia Poon. 2021. *Making Kin: Ecofeminist Essays from Singapore*. Singapore: Ethos Books.

Wang, Guifang, Li Huawei Yang Yang, Sándor Jombach, and Guohang Tian. 2019. '"City in the Park", Greenway Network Concept of High-Density Cities: Adaptation of Singapore Park Connector Network in Chinese Cities'. *Fábos Conference on Landscape and Greenway Planning* 6 (1). https://doi.org/10.7275/th5k-5d16.

Wee, Yeow Chin. 2015. 'Bukit Brown habitat can be recreated'. *The Straits Times*, 20 July 2015. https://www.straitstimes.com/forum/letters-on-the-web/bukit-brown-habitat-can-be-recreated.

Yeoh, Brenda S. A., and Theodora Lam. 2022. 'Managing the Non-Integration of Transient Migrant Workers: Urban Strategies of Enclavisation and Enclosure in Singapore'. *Urban Studies* 59 (16): 3292–311. https://doi.org/10.1177/00420980221114316.

Zaki, Sheikh Ahmad, Toh, Hai Jian, Yakub, Fitri, Ahmad Shakir Mohd Saudi, Jorge Alfredo Ardila-Rey, and Firdaus Muhammad-Sukki. 2020. 'Effects of Roadside Trees and Road Orientation on Thermal Environment in a Tropical City'. *Sustainability* 12 (3): 1053. https://doi.org/10.3390/su12031053.

Conclusion

Todd LeVasseur

We all have them: dreams of utopia. The better world we want to see. Where we want to live. Places that are full of justice, care, respect, safety, bodily security, intergenerational community, honesty and self-actualization without judgement or coercion. Places with clean air, clean water, nutritious and accessible food, surrounded by biodiversity and its continued evolutionary flourishing. Note that we have now firmly moved into dreams of ecotopia.

This book was motivated by such ecotopian dreams (Barnhill 2011): dreams social, and thus economic, and also and always, environmental. It is a book of imagined plural pathways – those paths shaped by decolonial praxis and theory, by insights from degrowth theory and by solidarity with feminist or queer speculative visions – towards strong sustainability in ecologies of alive places on a planet going through climate triage. It is a book written as a vibrant and vulnerable love letter informed by ecolinguistical theory and praxis to a better Singapore. A Singapore that has already come so far in creating places of justice, care, respect, safety, security and community; in stewarding and managing ecologies of alive places full of flora and fauna, both native to Singapore and that migrate through its boundaries. My co-authors and I have benefitted from these micro-glimpses of what a city of the future can and should offer: accessible and affordable public transit, well-designed public housing, clean and safe streets, numerous parks, copious numbers of forested trails, ethnic diversity and a government that recognizes untoward environmental (and climatic) trajectories. We celebrate these *beneficial* aspects of Singapore and the hard work of the countless governmental and civic leaders and workers, alike, who have laboured to make Singapore what it has become. We wish it more of the same in these successes, scaled up and outward, internally and in the region, but with a clear vision towards an intersectionally just climate-resilient future and a recognition that a 'properly pluralist approach [to sustainable living] should explore how moral futures are being made by multispecies projects of inheritance, adaptation, and world-making' (Jenkins 2023, 631).

Yet, given the ecosophy of this book, I recognize, as its editor and in writing its conclusion, that Singapore equally and sadly all too often lacks in care and community, and thus, in honest and critical reflection. As explored in Chapter 1, I too have seen Singapore's LGBQT++ communities systematically marginalized and heteronormativity dominate the housing market; I have seen the stress in Singapore's youth as they navigate a soul-smothering education system; I have seen the lack of resources for questioning and reshaping the too-often hegemonic 'Singapore Story'; I have seen contracted foreign labour work twelve-hour days, six days a week in trimming bushes, picking up trash and constructing boutique apartments; I have seen the last remaining Ubin Orang Pulau people constrained in their ability to practice their traditional culture on Pulau Ubin; and I have seen Singapore's last remaining forests and rivers and fields give way to reclaimed petroculture-based ecomodern urban fantasies that raze the ground and replace tropical soil with skyscrapers tinselled and *framed* with green vines that function as a veneer of sustainability.

Rather, the critical thrust of this book rests on seeing how 'the outlook of sustainability in Singapore today is decisively modern, technological, progressivist and highly pragmatic, and it needs to eventually relinquish this genealogy to better serve the environment' (Wee 2016, 66) and the various species present in Singapore's on- and off-shore boundaries. It also rests on the call from degrowth expert Raj Patel and the sociologist of capitalist world ecologies Jason Moore that what is needed is for us to 'develop a postcapitalist counter-hegemony' (2020, 205). That what is urgently needed in every urban area, including Singapore, according to them is 'the dramatic redistribution of resources that a movement beyond capitalism will require' (2020, 206). Thus, while the above sights I have experienced are in ways singular to Singapore, the trajectories are sadly all too real in urban areas worldwide. Or as the ethicist and political theorist David Naguib Pellow writes, 'Since most of the human beings on this planet are living in cities and other urban spaces, much of the future of … environmental justice struggles, and efforts to create resilience and sustainability will pay considerable attention to cities and the built environment' (2018, 20). Given this, then yes, some of the struggles and challenges on the path towards strong sustainability are and will be unique to Singapore and its own geopolitical past and present as well as its unique biogeochemical past and present. But many of the challenges – social, economic and environmental – are shared by humans everywhere. And they are particularly pronounced in urban areas as we approach and pass through urban metabolic resource bottlenecks and the planetary boundaries (as explained in the introduction) that will define the rest of our shared human lives in this century and beyond.

That these bottlenecks are coming to Singapore (and may already be at times present) is a given.[1] Singapore's own approach to its agri-food sector and its vision for Pulau Bukum – *destructive* discourses and 'texts' – inherently recognize this. That these highly resourced, well-meaning and intentioned visions cannot move beyond the ideologies of growthism and anthropocentrism attest to the point made by Patel and Moore shared above. The majority of Singapore's current vision for its sustainable future remains entirely tethered to these two *destructive* ideologies. Besides its ecomodern plans as articulated in the introduction, Chapter 1 and Chapter 3, in the years to come, Singapore also plans to:

- create via reclamation Long Island off of what is currently East Coast Park, thereby adding almost 2,000 acres and an eighteenth reservoir to Singapore's landholdings;[2]
- help construct a 2,600+ mile long undersea cable to receive renewable solar energy from Australia, while also getting renewable energy from neighbouring islands of Indonesia, with the latter potentially leading to the displacement of Indonesians on Batam and Rempang islands;[3]
- complete in the 2040s a mega-port (possibly the earth's largest port at the time of completion) at the cost of twenty billion Singaporean dollars and that can handle mega container ships and approximately sixty-five million containers a year, with much of this handled by automated shipping and artificial intelligence (AI) technology and built on reclaimed land.[4]

In these and similar plans, the ideologies of growthism and anthropocentrism that combine into a unique ecomodernism of Singapore continue unabated. Given the reasons provided in the introduction, Chapter 1 and Chapter 2, these are the same *destructive* and *ambivalent* discourses and their ideologies that lead to an ever-accelerating urban metabolism that must be challenged by Singaporeans. They are the same discourses that require the need for *beneficial* discourses like Chek Jawa, Bukit Brown Cemetery and the Park Connector Network. These and other *beneficial* discourses, texts, sites and the underlying ideologies they embody in Singapore can help challenge the reality as shared by Singaporean-born environmental historian Sunil Amrith that 'Singapore, an island-city that import[s] almost all its food and even its water, [is] a city as committed as anywhere on the planet to remaking nature for human ends' (2024, 1).

It is possible for new stories for Singaporeans to live by to emerge in the service of a post-capitalist, post-ecomodern ecotopia. *Beneficial* examples already exist, in the same geospatial contexts as Singapore's largely *destructive* ones, as for example in the difference between Pualu Bukom and Chek Jawa – both islands, but 'texts' that are shaped by entirely different ideologies. Or in the difference between a City in Nature discourse that perpetuates ecomanageralism and geo-power while commodifying nature in the name of 'sustainable development' (Luke 1995, 2005) with the discourse of the Parks Connector Network that instead helps create *community* and *care*. The latter are texts and discourses that align with this book's ecosophy in centring authentic and more-than-human community and its call to exhibit care for all members of such a community. They also align with this book's ecosophy and its critical reflection on Singapore's largely unfounded yet unabiding faith in growthism, human-centredness and ecomodernization as these all continue to threaten the planetary commons upon which Singapore itself depends for its own survival (Rockström 2024). They are *beneficial* texts and ideologies that reflect Stibbe's call for 're-writing and re-speaking the world' (2021, 211).

We hope that in your reading of this book, you have reflected on your own inherent earth wisdom and that you have built (or revisited) parts of your own ecosophy. That you have seen the value of *care* and *community*, and how we must all be critical consumers of any and all texts and discourses in this era of climate change and where we need to create non-hierarchical relations of collective continuance (Whyte 2018). We hope that you have been inspired to search for (or to create) new *stories to live by* that build on ideologies that can allow for earth flourishing.

And for those readers from Singapore, specifically, we hope you will join us in helping to re-write and re-speak what Singapore can and must become: a place of strong sustainability, of *care* and of *community*. We hope you actively find *salience* in Singapore's *beneficial* texts and its ecologies of alive places. And we hope you are *reminded* of your embodied and caring animal self, whether this reminding occurs in the *ambivalent* places highlighted in Chapter 4, or the *beneficial* places highlighted in Chapter 5, or elsewhere in the magical and special place we all know and love as Singapore.

Notes

1 And where such bottlenecks bely fantasies of 'green growth' (Kotz, Levermann and Wenz 2024). See also Lee, Xu and Kua (2023) and Liu and Christopolous (2024).
2 Gene and Begum (2023).
3 Jackson (2024) and Sud (2023).
4 Surbana Jurong (2021).

References

Amrith, Sunil. 2024. *The Burning Earth: A History*. New York: W.W. Norton and Company.

Barnhill, David. 2011. 'Conceiving Ecotopia'. *Journal for the Study of Religion, Nature and Culture* 5 (2): 126–144. https://doi.org/10.1558/jsrnc.v5i2.126.

Gene, Ng Keng, and Shabana Begum. 2023. 'Long Island to Be Reclaimed off East Coast Could Add 800ha of Land, Create Singapore's 18th Reservoir'. *The Straits Times*, 29 November 2023. https://www.straitstimes.com/singapore/long-island-to-be-reclaimed-off-east-coast-could-add-800ha-of-land-and-singapore-s-18th-reservoir.

Jackson, Lewis. 2024. 'Singapore Approves Import of Solar Energy from Australia via Undersea Cable'. *Reuters* (International), 22 October 2024.

Jenkins, Willis. 2023. 'Ethics After Humanity'. *Journal of Religious Ethics* 51 (4): 611–38.

Kotz, Maximilian, Anders Levermann and Leonie Wenz. 2024. 'The Economic Commitment of Climate Change'. *Nature* 628: 551–7. https://doi.org/10.1038/s41586-024-07219-0.

Lee, Sheryl Tian Tong, Selina Xu, and Isabel Kua. 2023. 'Singapore Is Fighting Rising Seas to Save $50 Billion in Real Estate'. *Bloomberg* (International), 9 October 2023. https://www.bloomberg.com/news/features/2023-10-09/singapore-is-saving-50-billion-in-real-estate-from-rising-seas-climate-change?embedded-checkout=true.

Liu, Jean, and Georgios Christopoulous. 2024. 'Commentary: A Hotter Singapore Will Change How We Work and Live'. *CNA (Asia)*, 22 April 2024. https://www.channelnewsasia.com/commentary/singapore-hot-weather-rest-break-aircon-night-4277606.

Luke, Timothy. 1995. 'On Environmentality: Geo-Power and Eco-Knowledge in the Discourses of Contemporary Environmentalism'. *Cultural Critique* 31: 57–81.

Luke, Timothy. 2005. 'Neither Sustainable nor Development: Reconsidering Sustainability in Development'. *Sustainable Development* 13: 228–38.

Patel, Raj, and JasonMoore. 2020. *The History of the World in Seven Cheap Things: A Guide to Capitalism, Nature, and the Future of the Planet*. Berkeley, CA: University of California Press.

Pellow, David Naguib. 2018. *What Is Critical Environmental Justice?* Medford, MA: Polity Press.

Rockström, Johan, Louis Kotzé, Svetlana Milutinović, Frank Biermann, Victor Brovkin, Jonathan Donges, Jonas Ebbesson, et al. 2024. 'The Planetary Commons: A New Paradigm for Safeguarding Earth-Regulating Systems in the Anthropocene'. *Proceedings of the National Academy of Sciences U.S.A.* 121 (5): e2301531121. https://doi.org/10.1073/pnas.2301531121.

Stibbe, Arran. 2021. *Ecolinguistics: Language, Ecology and the Stories We Live By*. Abingdon: Routledge.

Sud, Nikita. 2023. 'Victims of Green Energy Boom? The Indonesians Facing Eviction Over a China-Backed Plan to Turn Their Island into a Solar Panel "Ecocity"'. *The Conversation*, 23 October 2023. https://theconversation.com/victims-of-the-green-energy-boom-the-indonesians-facing-eviction-over-a-china-backed-plan-to-turn-their-island-into-a-solar-panel-ecocity-214755.

Surbana Jurong. 2021. 'Singapore's S$20-billion Tuas Terminal Takes Shape'. 21 December 2021. https://surbanajurong.com/sjconnects/singapores-s20-billion-tuas-terminal-takes-shape/.

Wee, H. Koon. 2016. 'Modernising Project of Sustainability: The Green and Blue Plans'. In *Singapore Dreaming: Managing Utopia*, 64–81. Asian Urban Lab.

Whyte, Kyle. 2018. 'Settler Colonialism, Ecology, and Environmental Justice'. *Environment and Society* 9: 125–44.

Index